Matthias Funk

Microstructural stability of nanostructured fcc metals during cyclic deformation and fatigue

**Schriftenreihe
des Instituts für Angewandte Materialien**

Band 9

Karlsruher Institut für Technologie (KIT)
Institut für Angewandte Materialien (IAM)

Eine Übersicht über alle bisher in dieser Schriftenreihe erschienenen Bände finden Sie am Ende des Buches.

Microstructural stability of nanostructured fcc metals during cyclic deformation and fatigue

by
Matthias Funk

Dissertation, Karlsruher Institut für Technologie (KIT)
Fakultät für Maschinenbau
Tag der mündlichen Prüfung: 11. Mai 2012

Impressum

Karlsruher Institut für Technologie (KIT)
KIT Scientific Publishing
Straße am Forum 2
D-76131 Karlsruhe
www.ksp.kit.edu

KIT – Universität des Landes Baden-Württemberg und
nationales Forschungszentrum in der Helmholtz-Gemeinschaft

KIT Scientific Publishing 2012
Print on Demand

ISSN 2192-9963
ISBN 978-3-86644-918-3

Microstructural stability of nanostructured fcc metals during cyclic deformation and fatigue

Zur Erlangung des akademischen Grades

Doktor der Ingenieurwissenschaften

der Fakultät für Maschinenbau
Karlsruher Institut für Technologie (KIT)

genehmigte

DISSERTATION

von

Dipl.-Ing. Matthias Friedrich Funk

aus Gunzenhausen

Tag der mündlichen Prüfung: 11. Mai 2012
Hauptberichter: Prof. Dr. Oliver Kraft
Mitberichter: Prof. Dr. Gerhard Dehm

Karlsruhe - 2012

Matthias Friedrich Funk

Microstructural stability of nanostructured fcc metals during cyclic deformation and fatigue

190 pages, 82 figures, 9 tables

Abstract

Nanostructured metals are fascinating alternatives for conventionally structured metals as they exhibit truly enhanced mechanical strength and hardness and therefore could allows to reduce weight and the use of resources. Nevertheless such suffer from microstructural degradation under mechanical or thermal load. Therefore, the focus of this work is the experimental observation of the microstructural stability of nanostructured metals under cyclic mechanical load. *Ex situ* tests of macro and micro samples were conducted to investigate the large scale behavior while *in situ* experiments in the TEM allowed to observe the nanostructural behavior and relate that to the dominant deformation mechanisms.

In *ex situ* fatigue experiments on nc Ni the microstructure coarsened and formed a bimodal grain size distribution in the vicinity of the crack tip. Almost no dislocations were observed in the fatigued samples. In terms of mechanical behavior, the nc Ni showed an enhanced fatigue lifetime compared to its coarse grained counterparts.

Similarily, nc Cu tested *in situ* in the TEM also formed a bimodal microstructure. The enhanced mechanical strength was shown to decrease as an impact of cyclic loading. In the vicinity of a growing crack, the grain boundary mobility on the nano scale seems to backup the Cahn-Mishin model of stress coupled grain boundary motion. The in situ tests also revealed how the crack tip can be arrested at coherent twin boundaries leading to a stepwise crack propagation. Behind the crack tip, bridging of "soft" grains appeared as crack opening continued.

The fatigue lifetime of nt Cu microsamples revealed a strong dependence on the applied frequency and is shifted to higher lifetimes with increasing frequency. The plastic deformation is characterized by a homogeneous deformation followed by softening and a shear band formation. The shear band features a rather shallow width of ~ 15 µm and undergoes large shear strains of up to 370%. The detected softening goes in hand with the decomposition of the nanotwinned structure and the formation of an ultra fine grained microstructure inside the shear band. The detwinning takes place by the motion of incoherent twin boundaries under an applied load.

The *in situ* deformation of nt Cu material reveals an ultimate strength of up to 2.5 GPa due to thin film effect and the small sample dimensions. In the nano samples, the decomposition of their nano structure by detwinning and twin thinning process was related to the motion of paired partials along coherent twin boundaries forming ledges. The observed shear fracture is in agreement with molecular dynamics simulations. Thereby, shear thinning is triggered by dislocation motion on {111} planes. In combination with the alternating twin modifications at high strain rates it leads to the observed zig-zag fracture.

Matthias Friedrich Funk

Mikrostrukturelle Stabilität von nanostrukturierten kfz Metallen während zyklischer Belastung und Ermüdung

190 Seiten, 82 Abbildungen, 9 Tabellen

Kurzzusammenfassung

Nanostrukturierte Metalle sind eine attraktive Alternative für konventionell strukturierte Metalle, da sie eine hohe mechanische Festigkeit und Härte besitzen. Dennoch kann sich ihre Mikrostruktur unter mechanischer und thermischer Last zersetzen. Deshalb liegt der Fokus dieser Arbeit auf der Untersuchung der mikrostrukturenllen Stabilität von nanostrukturierten Metallen unter zyklischer mechanischer Belastung. *Ex situ* Experimente wurden an Makro- und Mikroproben durchgeführt, um das globale Verhalten zu untersuchen. Andererseits erlaubten *in situ* Experimente im TEM eine Abbildung der Entwicklung der Nanostruktur und eine Korrelation mit den dominanten Verformungsmechanismen.

In *ex situ* Ermüdungsexperimenten an nanokristallinem Ni wurde eine Vergröberung und die Bildung einer Mikrostruktur mit bimodaler Korngrößenverteilung im Bereich um die Rißspitze beobachtet. In diesem Zusammenhang wurden in den ermüdeten Proben nahezu keine Versetzungen entdeckt. Im Bezug auf das mechanische Verhalten haben die nc Ni Proben im Vergleich zu grobkristallinen Metallen erhöhte Ermüdungslebensdauern.

In ähnlicher Weise bildeten auch die *in situ* getesteten nc Cu Proben eine bimodale Mikrostruktur. Es zeigt sich, dass sich die hohe Festigkeit unter monotoner Last durch eine zyklische Belastung im TEM erniedrigt. Im Bereich der sich ausbreitenden Rißspitze unterstützt die hohe Mobilität von Korngrenzen auf der Nanoebene das Model von Cahn-Mishin der spannungsinduzierten Korngrenzbewegung. In nanoverzwillingten Körnern stoppte der Riß an kohärenten Zwillingsgrenzen was zu einem stufenweisem Rißwachstum führt. Hinter der Rißspitze findet man einzelne weiche Körner die die sich öffnenden Rißflanken überbrücken.

Die Ermüdungsdauer von nanoverzwillingten Cu Mikroproben hängt stark von der Ermüdungsfrequenz ab und verschiebt sich bei höheren Frequenzen zu längeren Lebensdauern. Die dabei auftretende plastische Verformung ist durch eine homogene Verformung und anschließende Entfestigung und Scherbandbildung gekennzeichnet. Charakteristisch für das Scherband ist seine geringe Weite von ca. 15 µm und die hohe aufgeprägte Scherdehnung von bis zu 370%.

Die *in situ* Versuche an nanoverzwillingtem Cu zeigen aufgrund des Dünnschichteffektes und das kleine Probenvolumen eine Zugfestigkeit von bis zu 2.5 GPa. Der Prozess des Auflösen bzw. die Verschmälerung der Zwillinge während mechanischer Belastung wurde auf die Bewegung von Stufen in Form von gepaarte Partialversetzungen entlang von kohärenten Zwillingsgrenzen zurückgeführt. Das beobachtete Versagen durch Abscherung ist in Einklang mit Moleculardynamiksimulationen. Das damit einhergehende Ausdünnen durch Abscherung wird durch Versetzungsbewegung auf {111} Ebenen ausgelöst. Bei hohen Dehnraten führt diese Verformung, zusammen mit den sich abwechselnden Zwilllingsmodifikationen, zu dem beobachtetem zick-zack Bruch bei hohen Dehnraten.

Danksagung

Die vorliegende Arbeit wurde im Zeitraum von August 2008 bis März 2012 am Insitut für Angewandte Materialien am Karlsruher Institut für Technologie angefertigt.

Herrn Dr. Christoph Eberl gilt mein ganz besonderer Dank für das zur Verfügung gestellte Thema, die geduldige Betreuung und die ständige Motivierung und Diskussionsbereitschaft die mir sehr geholfen haben.

Prof. Oliver Kraft danke ich für die Übernahme des Hauptberichtes und anregenden Kommentare sowie Prof. Gerhard Dehm für die Übernahme des Nebenberichtes.

Dipl.-Phys. Aaron Weis, Dr. Christian Kübel, Dipl.-Ing. Ute Jaentsch, Dr. Torsten Scherer und Dipl-Ing. Daniela Exner danke ich für die Hilfe an TEM, EDX und FIB sowie Herr Dipl.-Ing. Ewald Ernst für die Hilfe bei Dreh-, Bohr- und Fräsarbeiten.

Prof. Daniel Kiener und Prof. Andrew Minor danke ich für die Betreuung und Ermöglichung der Auslandsaufenthalte in Leoben und Berkeley sowie Prof. Mughrabi und Prof. Peter Gumbsch für ausführliche Diskussionen.

Außerdem möchte ich mich bei meiner Zimmerkollegin Sofie Burger und allen anderen Kolleginnen und Kollegen des Insitutes IAM-WBM für das angenehme Arbeitsklima, gelegentliche Diskussionen und Beschäftigungen neben der Arbeitszeit bedanken.

Schließlich gilt mein Dank meinem Bruder Christoph und meinen Eltern Irmtraud und Martin für ihre Unterstützung jeglicher Art über all die Jahre hinweg.

Table of contents

List of symbols and abbreviations

a	length	(m)
A	area	(m^2)
$\lvert \vec{b} \rvert$	Burgers vector	(nm)
d	grain size	(nm)
D	diffuison constant	(m^2/s)
ε	strain	( )
E	Young's modulus / stiffness	(GPa)
G_{IC}	energy release rate	(J/m^2)
G	shear modulus	(GPa)
f	frequency	(Hz)
dE_{el}	elastically stored energy	(Nm)
dE	energy available for crack growth	(Nm)
dW	amount of inserted energy	(Nm)
h	film thickness	(nm)
K_{IC}	fracture toughness (valid)	(MPam$^{1/2}$)
K_Q	fracture toughness (not valid)	(MPam$^{1/2}$)
λ	twin thickness, twin boundary spacing	(nm)
σ_y	yield strength	(MPa)
R	stress ratio	-
UTS	ultimate tensile strength	(MPa)
U	acceleration voltage	(kV)

| U_{th} | beam damage threshold value | (keV) |

AFM   atomic force microscopy

BF   bright field

cg   coarse grained

CTB   coherent twin boundary

DB   domain boundary

DC   direct current

DIC   digital image correlation

DF   dark field

EBSD   electron backscatter diffraction

EELS   electron energy loss spectroscopy

FIB   focused ion beam

HCF   High cycle fatigue

HRTEM   high resolution TEM

ITB   incoherent twin boundary

LCF   low cycle fatigue

mc   microcrystalline

MEMS   micro electro-mechanical system

PSB   persistent slip band

nc   nanocrystalline

ns   nanostructured

nt   nanotwinned

PTP   push-to-pull

SEM   scanning electron microscopy

SF   stacking fault

SFE   stacking fault energy

SPD   severe plastic deformation

TBM   twin boundary motion

TEM   transmission electron microscopy

UFG   ultra fine grained

XRD   x-ray diffraction

1 Introduction

Nanostructured (ns) materials are solid materials with the smallest microstructural size between several nm and up to 100 nm. Ns materials have gained increased importance since the end of the 80s. This is a promising class of materials for versatile applications due to their enhanced properties in terms of strength, hardness and wear resistance, for example as hard corrosion and wear-resistive coatings. These materials can be employed to fulfill the technical requirements of future coatings for micromechanical and microelectronic applications (Figure 1-1). For example, ns Nickel-Tungsten can be used instead of toxic hexavalent chromium as a coating material. Besides the environmental advantages, it has a up to five times higher lifetime [1].

Figure 1-1: Ns materials are used e.g. for drill coatings [2], highly conductive Cu with enhanced strength in electronics [3], coatings in abrasive applications like backhoe bucket [4] as well as for optical coatings as a substitute when poisonous hexavalent chromium is to be prevented [5].

Generally, metals have high potentials, which can be exploited more effectively by a better-tuned microstructure. Since their production is still time consuming and expensive and also difficult, ns materials are often available only as thin films or foils. On the nanoscale mechanism come into effect which are remarkably less pronounced in the macro world. This can e.g. be due to the high surface-to-volume ratio of the tested volume, the large amount of interface, grain or twin boundaries per volume and the constrained dislocation motion. The influence of such effects on deformation mechanisms is investigated in this thesis. One of the main aspects of this work is the deformation behavior of ns materials under cyclic deformation and fatigue. Reasons for the incomplete understanding of these materials are challenges in their preparation, the often very small sample dimensions and sometimes their brittleness which involves special setups designed just for that purpose.

In the first section a literature overview on nanocrystalline (nc) and nanotwinned (nt) metals as well as some basics on fatigue of ufg and ns materials will be given. This is followed by chapters describing ex situ fatigue and in situ cyclic deformation of micro and nanosamples made from nc Ni and Cu as well as nt Cu.

2 Literature review

The modern beginning of nanotechnology dates back to the end of the 50s. First theoretical formulations originate from Richard P. Feynman [6] in his famous lecture "There is plenty room at the bottom" given at Caltech in 1959. He envisioned that the available space at small scales and dimensions is nearly endless. As the dimensions get closer to the size of the smallest atomic units, other forces, effects and laws are dominant compared to the ones we know from the macro world. This allows for new opportunities by the use of novel methods, technologies and new classes of materials.

In the late 80s first attempts to produce nanomaterials were driven e.g. by Herbert Gleiter [7]. Nanomaterials can be defined as materials with a smallest structural dimension of less than 100 nm. Material with a grain sizes between 100 and 1000 nm are called ultra fine grained (ufg) and above 1 µm is referred to as microcrystalline (mc) or coarse grained (cg) (Figure 2-1). An overview is given e.g. by reviews of Kumar and Suresh [8], Meyers et al. [9] and Koch [10]. The reviews mainly discuss nc Ni which was examined very well and is like an example material for face centered cubic (fcc) metals which are easily prepared by an electroplating process.

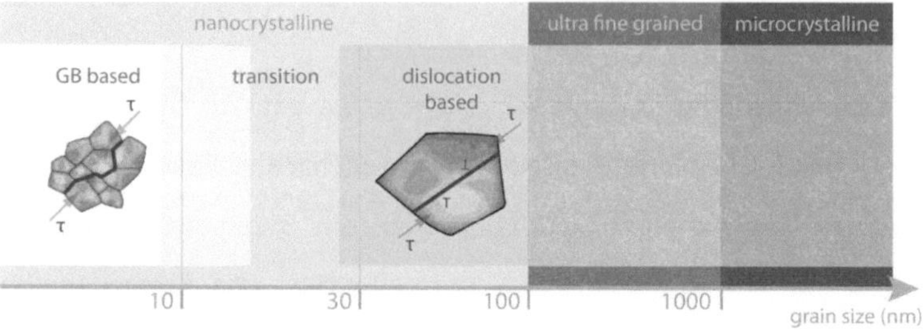

Figure 2-1: Regimes of nc, ultra fine grained (ufg) and microcrystalline (mc) materials. Deformation is dislocation based in the upper nc and grain boundary (GB) based in the lower nc region, with a transition in between. The red lines indicate the occurrence of deformation for example on the GB or within the grain.

The ns materials that are investigated in this work can be subdivided in nc and nt materials. Both are discussed further in the following sections.

2.1 Nanocrystalline materials

The fundamental processing routes for ns materials are the "bottom-up" and "top-down" principle. The first one works with the continuous deposition of single nanosized structures like grains and is basically used for thin films, as it is very time-consuming. Examples are magnetron sputtering and electrodeposition (ED). The sputter process is known for a low chemical contamination but fairly high porosity (in the range of up to a few nm sized pores) often found at grain boundaries (GBs). The ED-process provides a more dense material, but it usually comes at the cost of impurities especially on the GBs, e.g. sulfides on the GBs in nc Ni [11]. Complementary, there is the "top-down"-principle where the microstructure is being refined by mechanical severe plastic deformation (SPD) or similar techniques and introduction of a high dislocation density and subsequently formation of high angle GBs [12]. This comes at the penalty of a high defect density.

2.1.1 Mechanical properties of nc metals

Nc metals are interesting due to their increased hardness, strength and wear resistance with decreasing grain size. This is commonly known as the Hall-Petch relation, first described by Petch and Hall [13, 14] and extended by Petch [15]:

$$\sigma_y = \sigma_0 + kd^{-1/2} \tag{2-1}$$

Here σ_y and σ_0 are the yield strength of the nc and the conventional material, d is the grain size and k is a material specific factor (see Figure 2-2).

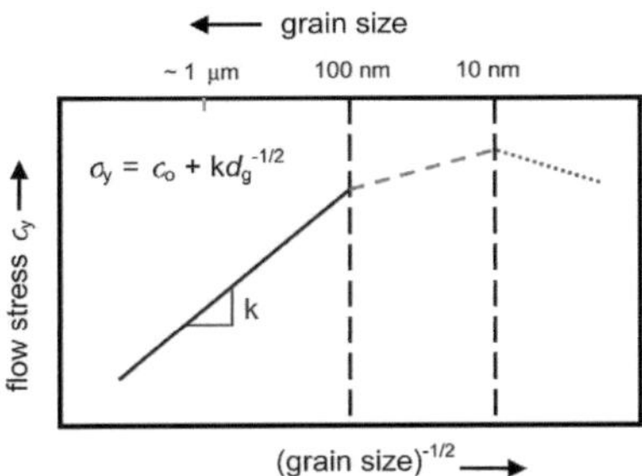

Figure 2-2: Schematic curve that illustrates the Hall-Petch relation with increasing strength down to 10 nm, followed by the Hall-Petch breakdown [8].

The increase in strength is induced by a pile up of dislocations at GBs and the resulting back stress on the dislocation source. The smaller the grains become, the smaller the mean free path of the dislocation will be until they reach the GBs. In grains smaller than ~ 10 nm, this trend changes and is reversed or reaches a plateau, due to a change in deformation mechanism (see later chapter). This is known as Hall-Petch breakdown [16] [17] [18] [19] or inverse Hall-Petch relation [20, 21]. Below grain sizes of 100 nm there are not enough dislocations able to cross the grains to maintain significant Hall-Petch hardening. In this regime, the so called partial dislocation extension model [22, 23] is expected to lead to an increase in strength and redeem the Hall-Petch-hardening.

2.1.2 Scaling- and size effects

There are two effects that influence the material strength depending on structural dimension. First is the scaling effect which is purely geometrical. If samples get smaller, the surface-to-volume-ratio increases (Figure 2-3). The more surface the more sites for crack formation increase crack probability. This appears in addition to size effects, where the mechanical behavior is depending on a characteristic intrinsic or extrinsic dimension. The classical size effect is often correlated to the defect distribution which can be related to the manufacturing process (e.g. pore size distribution in molded samples). Similarly the fatigue strength increases when the sample volume is reduced, that is under high stress during the test [24]. The behavior can then be described by a stochastic behavior e.g. by the Weibull-distribution, where the strength of a material is defined by the biggest defects, which are inherently distributed. When the size of the tested material is reduced, the distribution changes and the probability to have a defect which leads to fatal failure at a given stress is smaller, thus increasing the strength [25] (Figure 2-3, right). Secondly, in smaller samples the stress gradient and the stressed material volume are large [26].

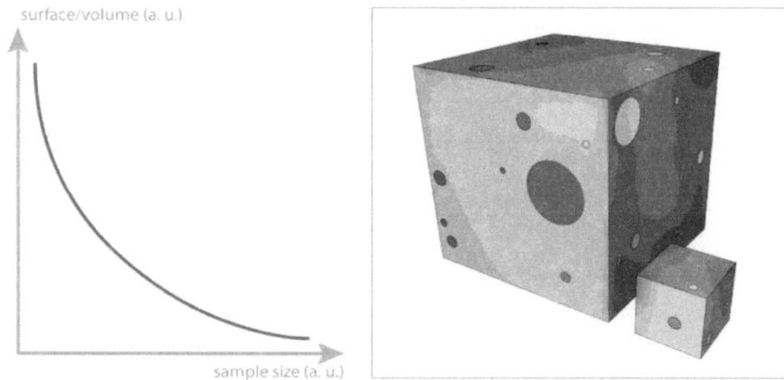

Figure 2-3: Schematic increase of surface-to-volume-ratio, with decreasing grain size (left), schematic of a microstructure with pores as defects distributed over the hole sample a bigger defect is included as in the small section (right).

When the smallest dimension of a sample is reduced to a point where the governing mechanisms cannot freely act to carry deformation, the strength of the material is influenced by the interaction of the mechanism with this dimensional constrain. This can even lead to a complete change of the active deformation mechanism. With nanoindentation it was clearly shown that the strength or hardness increases when the tested volume gets smaller [27].

2.1.3   Grain boundaries

As the content of GBs increases in finer grained materials, its importance for the overall mechanical behavior increases. In early publications on nc materials, GBs were regarded to be amorphous [7]. At a grain size below around 3 nm even the whole material was regarded to be in an amorphous state. This model was revoked in later publications [28, 29], where GBs were described as a region with a clearly defined structure (Figure 2-4 left)[[30].

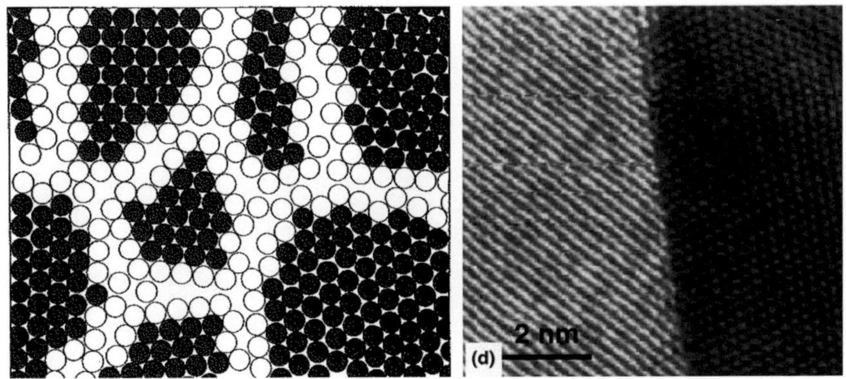

Figure 2-4: Schematic image of the grain structure of a nc material with grains (black) and "amorphous" GBs (white) (left) [7]. In a more recent model the GBs do not consist of something like a amorphous phase any more (right) [9].

Even though, GB regions are not that periodic and defined in terms of structure as the grain interior. Therefore they are not as closely packed and the additional space that results from this is called the "excess volume". As a consequence the mechanical properties as well as the diffusivity along the GBs might be affected. The excess volume is larger in nc materials as in their coarse grained counterparts [31]. While grain coarsening is occurring, the excess volume may be passed over into excess vacancies which can be annihilated (in vacancy sinks or at the surface) or can form pores. This can be a large effect in thin films.

2.1.4   Dislocation constraints and the impact on ductility

Along with the high strength and hardness in nc materials comes reduced ductility. This is based e.g. on the lack of dislocation motion in nano sized grains. Dislocations are responsible for the accommodation of plastic deformation and reduced stresses in coarse grained materials. Additionally in the nano sized grains, the image forces on dislocations are high. Therefore, the dislocations are pulled to GBs and become incorporated. In the nc metals it would take too much force to get the dislocations dissolved from the GBs into the grains.

Further sources for brittle behavior in nc materials are:

- Porosity, which is typical for sputter deposited materials, which for thicker films like coatings can be improved by ball milling [32]
- Contaminations, for example sulfides, which are typical for electrodeposited materials (sulfates are added to the electrochemical bath in order to control the grain size) [33]
- Strain localization, due to the nano sized grains with a reduced dislocation density and motion and therefore suppressed strain hardening.

2.1.5 Instability – ways to increase ductility

The limited ability for strain hardening is a major reason for the low ductility of nc materials. Coming from larger grains, the materials capability for strain hardening is strongly bound to the dislocation density (ability of dislocation multiplication / emission) and described by [26]:

$$\Delta\sigma = \sigma_i + \alpha Gb\rho^{1/2} \qquad (2\text{-}2)$$

With $\Delta\sigma$ and σ_i as the increased and initial strength, α is a factor, G the shear modulus ρ the dislocation density. Consequently, when a defect is appearing at the surface, the material can't strengthen at that region in order to stabilize it by shifting the plastic deformation to another region. That would happen in coarse grained materials by strain hardening:

$$\sigma = K\varepsilon^n \qquad (2\text{-}3)$$

σ and ε are stress and strain, K is the material specific strength coefficient and n is the strain hardening exponent defined by (Consideré criterion)

$$n = \frac{\varepsilon}{\sigma}\frac{d\sigma}{d\varepsilon} \qquad (2\text{-}4)$$

Nevertheless, mechanical strain hardening was observed in ED nc Ni [34]. Hanlon et al. refers this behavior to dislocation source exhaustion during cyclical loading and the need for higher stress to activate another set of dislocation sources. A more practical way to enable strain hardening is to adjust a bimodal microstructure [35]. The small grains can counter for the high mechanical strength and the larger grains can carry the deformation and accommodate stresses. Even though, this would theoretically mean a tradeoff with a lower strength than pure nc metals [36, 37, 38]. Materials with a bimodal microstructure can promote high mechanical toughness and introduce ductility and therefore have increased fracture toughness.

2.1.6 Strain rate sensitivity and activation volume

The strain rate sensitivity and activation volume are two parameters to characterize a material in terms of correlation between dislocation motion and externally applied load [39]. The stress σ is correlated to the strain rate $\dot{\varepsilon}$ by:

$$\sigma = k_s \dot{\varepsilon}^m \tag{2-5}$$

With k_s being a prefactor, and m is defined as the strain rate sensitivity:

$$m = \frac{\partial \ln(\sigma)}{\partial \ln(\dot{\varepsilon})} \tag{2-6}$$

As the coarse grained materials deform mainly by dislocation motion, these are relatively strain rate independent. When the grain size is reduced, the content of deformation that is correlated to diffusive processes increases. The average volume involved in the deformation process is called the activation volume V after [40, 41]:

$$V = 3\sqrt{3}kT\frac{\partial ln\dot{\varepsilon}}{\partial H} \tag{2-7}$$

k is Boltzmann's constant, T the temperature and H the hardness. The strain rate sensitivity is increasing and the activation volume is reduced and finds its minimum between a grain size of 10 and 20 nm. This is the regime, where the transition of deformation mechanism from dislocation based deformation to GB mediated plasticity is being expected [16] (see Figure 2-1).

2.2 Nanotwinned materials

Nanotwinned (nt) materials feature high mechanical strength and a high electric conductivity [42] as well as enhanced mechanical ductility [43]. This form of materials was first produced by Dahlgren [44] in a sputter process resulting in homogenously oriented twins. Alternatively they can be produced by electro deposition (ED) where the orientation of the twins is random [45]. During the magnetron sputter process of Cu (111) planes on HF etched Si (110), the atoms of the 1st layer start deposit as islands in one possible orientation (A). The 2nd layer has two possibilities, to deposit in configuration B or C (see Figure 2-5 a). The stacking sequence is more or less arbitrary and can be different in the different domains. These domains are separated by incoherent twin boundaries (ITB) which form when the islands grow together. When the stacking sequence changes within one domain, a coherent twin boundary (CTB) forms (see Figure 2-5 b) [46].

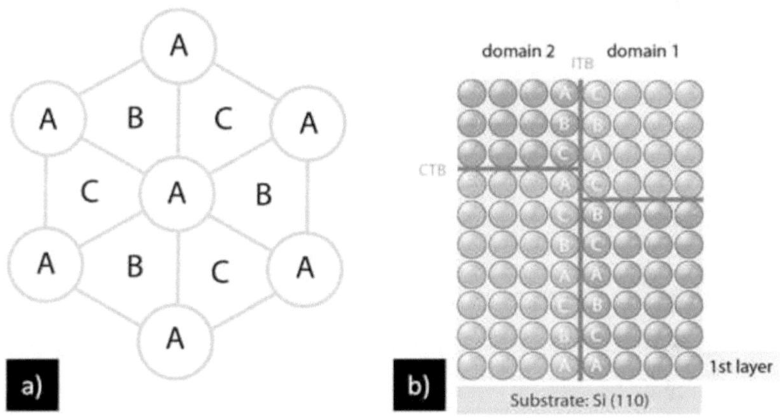

Figure 2-5: (a) Schematic drawing of stacking sequence in a fcc structure from top view. (b) schematic of a side view of two domains, containing CTB and ITB.

To form the desired size of the growth twins, a material with a low stacking fault energy (SFE) and twin boundary energy (TBE) is sputtered at high deposition rates [43, 45]. There are a few fcc metals that have a lower SFE and TBE than Cu (45-80 and 24-39 mJ/m^2), for example Ag (22 and 11 mJ/m^2). Ag tends to be very "unclean" in the sputter process and tends to contaminate the sputtering chamber. The microstructure of nt materials usually consist of an ufg or even coarser microstructure with ultra fine domains with a size of 200 – 600 nm. Each single grain is interspersed with twins of a thickness in the nano regime. When the domains are oriented with almost no misorientation, the microstructure can be regarded as single crystal with domains of stacked nanotwins [45].

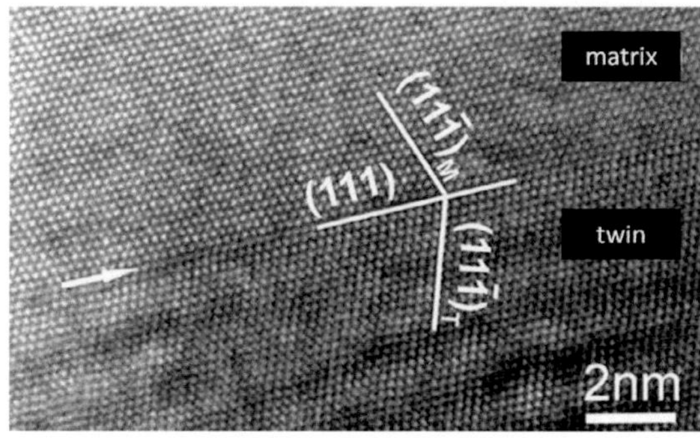

Figure 2-6: HRTEM micrograph of a twinned structure with twins between a CTB (adapted from [47]).

2.2.1 Mechanical properties

The twinning mechanism is not only used to explain deformation, but also to systematically produce materials with high strength and high ductility, which can be increased even more with decreasing grain size [8, 45, 48]. Further advantages of nt metals, besides the higher ductility compared to nc metals at the same stress levels, are the higher thermal stability of the microstructure since TBs have a lower energy as GBs [49, 50, 51]. These boundaries represent a similar barrier to dislocation motion as GBs due to the discontinuity of the slip systems [52]. Consequently the Hall-Petch relation works as well. For example, after annealing up to 800°C the twins do not grow more than 20 nm [53]. Nt materials have a promising combination of strength and ductility in combination with a fracture initiation tolerance and damage tolerance leading to high toughness values. In contrast, in microcrystalline (mc) materials, a reduction of the structural size usually leads to a reduction of fracture toughness; whereas a reduction of twin thickness leads to an increase in toughness [54]. Twin boundaries with spacing down to 30 nm can act like dislocation sources but also as obstacles (improved ductility and strength at the same time). Below 30 nm a transition occurs and further increase of the yield stress leads to the emission of partial dislocations [55]. Similar to nc materials in nt metals a maximum of strengthening appears at ~ 10 nm twin thickness [43], which also fits into the Hall-Petch model for GB strengthening [56]. This regime is followed by a plateau or even decrease of strength, as well as the motion and annihilation of twins by partial dislocation motion [57].

2.2.2 Dislocation twin interaction

The interactions of dislocations and TBs have been discussed for a long time. However, the exact mechanisms stayed unknown [58] until MD simulations gave [59, 60] better insights of the underlying processes. Similar to GB in nc materials, TB in nt materials have effective mechanisms to impede dislocation motion. The dislocations seem to be blocked due to the discontinuity (mirror symmetry) of the slip systems. As a consequence, the partials cannot just pass the TB, but have to change order, which again needs some energy. Locks can be formed at the TB, which act as another strengthening mechanism. In nt materials, with its ultra fine grains or domains, dislocation formation and motion is generally not suppressed, at least in direction normal to the CTBs. In the initial dislocation free microstructure they can be emitted from GBs or ITBs. Once these dislocations reach the TB, they can react which can lead to the nucleation of Shockley partials [61]. This was detected in post-mortem TEM observations [62] where a significant amount of dislocations debris was found around GBs and TBs after mechanical loading. The key advantage of nt materials is that they have big enough domains to allow dislocation motion and multiplication. Additionally they have TBs, which act as dislocation barriers and sources and allow hardening and provide additional dislocations.

Possible constellations of Σ3 {111} CTBs and partial dislocations are illustrated in Figure 2-7 [56]. With a single dislocation remaining in a twin, the TBs are not affected (a). With a twin dislocation moving along a twin boundary, the twin can become thicker or thinner, but the TB area remains the same (b), even with multiple dislocations involved (c). In case (d) an arrangement of dislocations forming an ITB segment is converting the twin into the matrix orientation. This process is energetically more favorable because the TB area is reduced. From a kinetic point of view, this process gets more difficult for thicker twins. As more twin dislocations have to be involved, therefore the process becomes more un-likely and therefore thinner twins are more prone to dissolve[55].

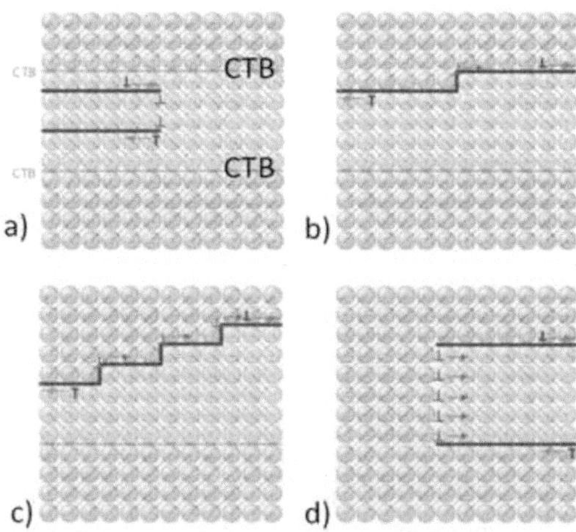

Figure 2-7: Schematics of the different dislocation twin boundary interactions that could lead to detwinning. Ether in a stepwise thinning (b,c) or a collective motion of an line of dislocations over the whole length of a twin (adapted from [56])

Deformation twinning can be regarded as an alternative deformation mechanism, when dislocation based deformation is suppressed like in many bcc metals or in fcc metals at low temperatures and high strain rates [63]. Reaction of dislocation and TBs can lead to dissociation of the dislocation and a step in the twin which can be regarded as twin boundary migration [64]. Even though this process is less likely than GB motion due to the excess energy of TB being lower [43, 56].

2.3 Deformation mechanisms and processes

As shown in Figure 2-1 and [65] the nc regime can be divided into three regions:

- < 10 nm: GB related processes are the dominant deformation mechanisms

- 10 - 30 nm: crossover regime, continuous transition from intragranular dislocation to intergranular GB plasticity, while partial dislocations are emitted from GBs

- 30 - 100 nm: dislocation-mediated plasticity is dominant, like in cg materials

The upper limit of the transition region is discussed to be in the range of 50 nm in [66]. At smaller grain sizes, GB mediated processes become much more likely as the content is increased [7] and dislocation mediated deformation is suppressed [65, 67]. Alternative deformation mechanisms have to be active like GB sliding and rotation [9, 30, 68, 69, 70]. Even though the transition between the different regimes is not sudden.

A good overview over the possible constellations between grain size and applied stress is given in so called deformation mechanism maps [71]. A central parameter is the stacking fault energy (SFE) and the elastic constants. For example, in metals with a low SFE energy, a second partial can be emitted to form a perfect dislocation. In the low SFE metal, the emission of a second partial from the GB is suppressed and the first partial would move through the whole grain. As it can block the propagation of any later-formed partial it acts as a mechanism for strain hardening.

The deformation mechanisms and processes can be classified in their dimensionality. If the dominant aspect of the deformation is 0-, 1- or 2 dimensional:

0-dimensional: Diffusion based

1-dimensional: Dislocation based

- Conventional dislocation based plasticity
- Partials: twins (same atomic plane), SF (different atomic plane)

2-dimensional: GB based

- Normal motion (diffusion induced GB migration)
- Relative translation (coupled GB motion)
- Rigid body translation (GB sliding) & grain rotation

2.3.1 Diffusion based deformation

Diffusional mass transport correlated to deformation and microstructural changes might play a significant role in nc materials. The increased excess volume along the GB provides paths of preferred and faciliated diffusion. The diffusion coefficient D is not enhanced per se, for materials with higher GB content like in nc Cu (IGC, which is similar to sputtered Cu) compared to cg Cu [72]. Although, the higher GB content implies shorter diffusion distances between triple points. The time dependent correlation between diffusional length x_D and the time t_f is given by [73]:

$$x_D \sim \sqrt{6 * D * t_f} \qquad (2\text{-}8)$$

For example the time needed for an atom to diffuse 1 µm along a GB is 10000 times longer than it needs for 10 nm. In conditions with significant diffusion, additional application of mechanical load is causing creep processes. Three different types of creep can be considered [74]:

- Coble-creep is related to diffusive processes along the GB. The dependence is $\dot{\varepsilon} \sim d^3$ therefore it is important at small grain sizes, even though it might be negligible at room temperature [75]. For example, D in Cu is increased by a factor of 10^9 at 8 nm Cu compared to lattice diffusion (293 bis 393 K) due to short circuit diffusion paths [76]:

$$\dot{\varepsilon} = A_C \frac{\sigma \Omega}{kT} \frac{\delta D_{GB}(T)}{d^3} \qquad (2\text{-}9)$$

- Nabarro-Herring-creep is related to volume diffusion and the dependence is $\dot{\varepsilon} \sim d^2$. It is not expected to play a significant role in nc materials:

$$\dot{\varepsilon} = A_{NH} \frac{\sigma \Omega}{kT} \frac{D_V(T)}{d^2} \qquad (2\text{-}10)$$

- At higher temperatures, grains can slide on each other. This is called GB sliding. At lower temperatures it could play an important role for small grain sizes, as it ensures the compatibility between grains. The strain rate is given by:

$$\dot{\varepsilon} = A_{GBS} \frac{\delta \sigma^n D_{GB}(T)}{d} \qquad (2\text{-}11)$$

2.3.2 Dislocation based deformation

Dislocation based deformation is regarded to be the classical mechanism that allows plasticity in conventional materials. It might be active down to 30 – 50 nm, with decreasing effectiveness, as the dislocation density is decreasing with decreasing grain size. Forces to activate dislocation sources (Orowan mechanism) within grains become too high at this scale and the only dislocations that appear are emitted from GBs. The shear stress to bend a dislocation in a material with shear modulus G and Burgers vector b to a radius R is approximately:

$$\tau \approx \frac{Gb}{2R} \qquad (2\text{-}12)$$

This means, that the shear stress to force a dislocation to bend in a single grain is increasing proportional to the decreasing grain size. For Cu with G $\cong$ 48 GPa, a grain size of 50 nm and a Burgers vector of b $\cong$ 2.5 Å, a minimum shear stress of 120 MPa is needed. Accordingly, a shear stress of $\tau \cong$ 1.2 GPa is needed for classical dislocation motion through 5 nm grains.

2.3.3 Partial dislocation based deformation

As in nc materials the grain sizes are getting smaller, the forces, needed to activate dislocation motion in grains becomes larger. Therefore, it is more likely and energetically favorable that dislocations split up and form partials which require less energy [9]. As GBs are sites of increased energy, they can act as partial dislocation sources and sinks at the same time. This was observed experimentally [77] as well as in simulations [78]. A shear stress is required to emit partial dislocations from the GB, or a stress concentration at the boundary, to generate a stacking fault (SF) between the leading and the trailing partial. The leading partial would progress through the grain and is absorbed in the GB on the opposite side of the grain. The trailing partial could follow and neutralize the SF produced by the first one. This process leads to stacking faults or twins. The partial dislocation line is most likely not parallel to the GB or the original dislocation line; therefore, this model is called the non-uniform partial dislocation extension model. The necessary shear stress to emit the partial is inversely proportional to the grain size [22, 23]. This model was later generalized to describe hardening by GBs in nc and by TB in nt materials [23]. The whole process is strongly depending on the SFE below ~ 30 nm grain size in nc materials [79] and twin spacing in nt materials [55]. The correlation of the grain size d, the twin spacing λ and the necessary shear stress to emit the partial is given by [40]:

$$\frac{\tau}{G} = \frac{\Gamma}{Gb} + \frac{1}{2}\left(\frac{1}{3} - \frac{1}{12\pi\beta_1}\right)\frac{b}{\lambda_1} + \frac{1}{2}\left(\frac{1}{3} - \frac{1}{12\pi\beta_2}\right)\frac{b}{d} \qquad (2\text{-}13)$$

Where G is the shear modulus, b the Burgers vector, λ the twin spacing, d the grain size and Γ and β are constants. The dependence of the strain rate sensitivity together with activation volume of grain size was also presented with the partial emission and extension model [40]. The motion of twin boundaries (TBM) by emission can be explained by the motion of Shockley partials [47].

2.3.4   Grain boundary migration

GB migration is not a deformation mechanism per se as the migration is not coupled to a plastic deformation. GB configurations try to reach an equilibrium state with the lowest possible energy available but the diffusion based process is thermally activated. Therefore kinetics is limited at RT or lower temperatures. Capillarity forces due to GB curvature are the driving force in GB motion; whereas, a small grain size with a high curvature results in a large driving force. To maintain the 120° angle at triple points, which represents the state of lowest energy, the triple points have to move as well. The driving force F_i, normal to the GB surface is inversely proportional to the GB curvature. Thus it is high for small grains and becomes lower as grains grow [80].

2.3.5 Relative translation coupled to normal GB motion

When an external stress is applied, the GBs try to reach an equilibrium state similar to the case mentioned above. In contrast, the stress coupled GB motion happens at a higher speed as they are not diffusion bound. Symmetrical high angle GBs can be regarded similar to low angle GBs by an array of edge dislocations. It also leads to shear, parallel to the GB plane, when the dislocations move in this glide plane parallel to the normal vector of the GB. As already mentioned, the mechanism is not based on diffusion. Atoms are 'transferred' from grain 1 to grain 2 as a whole cell of atoms, which is called a "raft". This raft rotates from the configuration A into configuration B (see Figure 2-8). In doing so, the GB is shifted by one layer. Problems that could arise with coupling in polycrystals are orientations of neighboring grains that are random. Consequently, the coupling factors must be compatible. GB curvature can disturb the coupling and the coupling factors have to be compatible [80]. Overall it is a mechanism that is likely in nc materials due to the smaller GBs and the higher stresses while it is connected to grain rotation and grain growth [30].

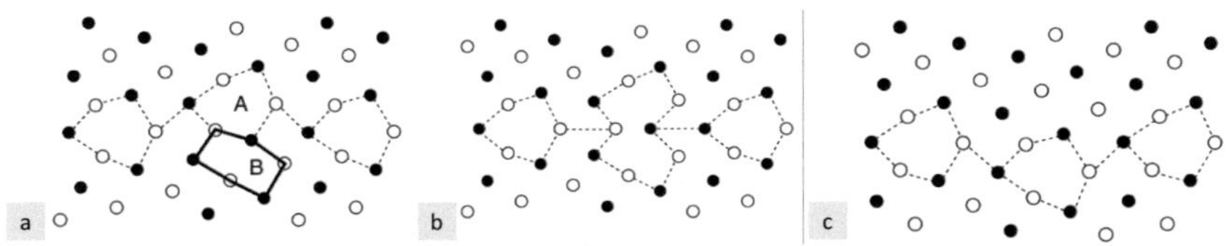

Figure 2-8: Schematic illustration of the stress coupled GB motion. The "atomic cell" labeled B in the left image rotates by minimal shift of single atoms. In image b and c the whole cell rotates further and the GB has shifted (adapted from [30]).

2.3.6 Grain boundary related process; GB sliding and grain rotation

At the smallest grain sizes, grains could glide off along GBs [66]. This happens easier, the less the grains are interconnected to each other and the smaller the grains are. Always in combination with GB sliding is grain rotation. The smaller the grain size gets, the higher the probability for rotation. It scales with d^{-4} with d being the grain diameter [67, 65]. During deformation grains rotate in each other's orientation and once they are equally aligned they can appear as one conglomerate (see Figure 2-9). This can then be big enough to carry dislocation motion and conventional deformation or fatigue mechanisms. This process is commonly known as coalescence or agglomeration [9]. Critical is the fact that there might be some offset between the glide planes between neighboring grains. Depending on how big this offset is, dislocation motion from one grain into another might be prevented, just like at regular GBs. When grains are still in connection with surrounding neighbors dislocations can be created [81].

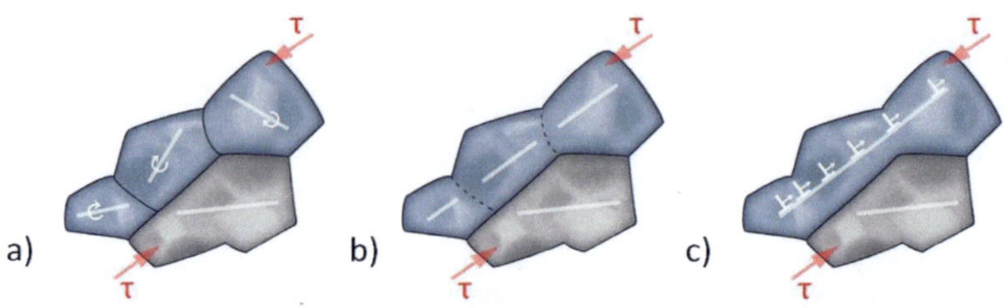

Figure 2-9: Schematic illustration of the grain rotation process. (a) Randomly oriented grains rotate into each other's orientation until they are aligned (b). (c) As they have no more different orientations, the grains coalescence and dislocation motion over several "former" grains is possible.

In materials with a monomodal nc grain size distribution, stress concentration could appear at GB triple junctions during mechanical loading. As a consequence, GB dislocations are nucleated, which tend to climb along the GBs. This process invokes in a further step the grain rotation. The rate is depending on the GB diffusion. Alternatively, in a bimodal nc structure, lattice dislocations are formed in a 1st step by plastic slip deformation in big and therefore "soft" grains. The pile-up of those dislocations acts as a stress-concentrator, similar as triple junctions. The rotational process was simulated in MD simulations by [82] and experimentally shown on nc Ni in-situ TEM [67]. The grain rotation is enhanced by high stresses, for example at a crack tip.

2.4 Fatigue / cyclic loading

Fatigue is called the microstructural change or damage during cyclical mechanical loading below the ultimate strength of a material. A convenient way to visualize the results from fatigue experiments is the S-N-curve (Woehler-plot) where the stress amplitude is plotted versus the number of cycles to failure N_f in a double-logarithmic manner [83]. Alternatively the strain amplitude ε_A can be plotted in a Coffin-Manson plot. In the following sections the different regimes of fatigue, the conventional failure process, the control of the experiments as well as the possible microstructural evolution are introduced.

2.4.1 Regimes of fatigue

The S-N-plot consists of a low cycle fatigue - LCF and a high cycle fatigue -HCF regime. In the LCF regime, described by the Coffin-Manson-law, higher loading into the plastic deformation damages more of the material leading to a lower number of cycles and the lifetime is typically limited by crack propagation. In the HCF regime, described by the Basquin-law, the deformation is mainly elastic (apart from microplasticity, which leads to crack formation in the HCF regime) a higher number of cycles to failure are measured [84]. The combination of both regimes leads to generalized formulation which describes the whole range of loading in terms of cycles and load (see Figure 2-10).

Coffin-Manson (LCF):

$$\varepsilon_A^{(pl)} = \varepsilon_f'(2N_f)^{-b} \tag{2-14}$$

Basquin (HCF):

$$\varepsilon_A^{(el)} = \frac{\sigma_f'}{E}(2N_f)^{-a} \tag{2-15}$$

Sum:

$$\varepsilon_A = \varepsilon_A^{(el)} + \varepsilon_A^{(pl)} = \frac{\sigma_f'}{E}(2N_f)^{-a} + \varepsilon_f'(2N_f)^{-b} \tag{2-16}$$

Where E is the Young's modulus, ε_f is a benchmark for the true strain to failure and σ_f is a material constant closely related to the ultimate strength. a is the fatigue strength exponent, which is a material property and depends also on the geometry and is usually between 0.05 and 0.12. b is the ductility exponent and depends on the strain hardening with values usually between 0.4 and 0.73 [74].

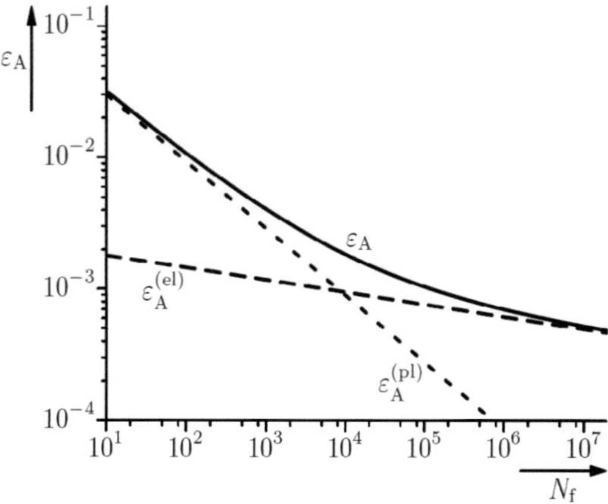

Figure 2-10: Fatigue life diagram with elastic (Coffin-Manson) and plastic (Basquin) regime [74]. The region in the mainly plastic regime is called LCF and in the elastic regime it is called HCF ~ > 10^4 cycles.

2.4.2 Crack initiation – PSB- Crack growth and closure

During fatigue of conventional grained materials, dislocation motion leads to extrusions and intrusions, which lead to a roughening of the surface. In combination with oxidation of those newly created surfaces, sharp cracks are formed where stress concentrations develop at the surface. Slip bands form at the surface as a result of persistent slip bands (PSB) in the material, which are preferred paths for the dislocations' back and forth motion. On the surface, these PSBs appear in form of extrusions. The largest amount of time is needed for crack formation. Therefore most of the mechanisms to suppress fatigue aim on the suppression of the crack formation [85]. That process is impeded in nc metals (higher strength) in the elastic HCF regime which results in a higher fatigue strength. The common idea is, that in nc materials it is harder to form cracks on the surface after the extrusion-formation mechanisms [86]. The reason is the lack of available dislocations for the formation of extrusions.

Once a crack is formed at a stress concentration it can propagate very easily through the material as no crack deflection is observed e.g. in nc Ni [87] [88]. This would consume energy and slow down the crack as crack tortuosity plays an important role during fracture or crack growth in coarse grained metals. In contrast, in case of mc or ufg materials the cracks are forced to follow the GBs on an energy consumptive path.

2.4.3 Control of the fatigue experiments

In principle there are two ways to run a fatigue test:

- Load controlled or
- Displacement controlled.

Accordingly, a fatigue-diagram can be created with a stress or a strain axis as an ordinate, correspondingly a Woehler-plot, or a Coffin-Manson plot. In a load controlled fatigue test the load signal of the load cell is used to control the displacement, to keep the load on the sample in the required limits (for example between 10 and 30 N, which can correspond to a stress of 100 and 300 MPa). When a crack grows, the cross section of the sample is reducing and thus the stress is increasing (load stays within the preset limits) and the crack is growing even faster.

In a displacement controlled fatigue test the crosshead of the testing machine is moving between two reversal points of displacement x_1 and x_2 at a specific strain rate or frequency. Therefore it is not really controlled with a feedback signal of the sample. During a fatigue test, especially in LCF, the sample would extend plastically. As the reversal points of the crossheads stay the same and are set for the initial, not elongated state of the sample, the stress corresponding to both reversal point are steadily reduced and could even go into compression. As no control loop is necessary, displacement control is more feasible. The advantage for displacement controlled fatigue testing is that in case of crack growth, the stress is not increasing due to the reduction in cross section. Therefore, the crack could propagate in a more restrained manner. In case of load control setting, sudden failure is more likely once a crack appears.

2.4.4 Microstructural evolution

Depending on the initial grain size and microstructure, materials may react with a different development of the microstructure during fatigue. Coarse grained materials are likely to form dislocation networks like cells, walls and persistent slip bands (PSB) [89, 84] and in general increase their dislocation density. In materials with an initially small grain size or high dislocation density, the microstructure tends to annihilate dislocations, decrease the dislocation density and increase the grain size [90] [91], leading to softening. As fatigue failure usually starts at the surface, coarsening has a large effect on the fatigue properties. For example, residual stresses like compressive stresses, increase fatigue life and tensile stresses decrease it [85]. But also more recent studies by Moser and Hanlon showed cyclical hardening during fatigue of nc materials [92]. In that case, after exhaustion of dislocation sources (which are usually located in the GBs) the stress has to be increased to activate another set of sources on other glide planes. Therefore, the strength is increasing which can be regarded as cyclic hardening. The mechanisms for coarsening, whether it is conventional coarsening or coalescence/rotation are still under investigation [9]. The fatigue in nc metals was reviewed in detail by Padilla and Boyce [85] and in terms of fatigue of small samples by Hanlon et al. [88] and Conelly et al. [93].

2.4.5 *In situ* testing in transmission electron microscopes

In mechanical testing it is often even more interesting to be able to observe the sample and microstructural details during mechanical loading. Therefore, efforts have been made to develop systems capable of *in situ* testing since decades. Since about ten years, techniques are available to accurately measure the load on those nanoscaled samples during testing in detail and to gather qualitative data at the same time. A large step in that progress was the development of micro electro mechanical systems (MEMS) for the use with *in situ* TEM measurements e.g. by Haque and Saif [94, 95, 96] or more recently by Espinosa and Zhu et al. [97, 98]. Proper stress measurement in combination of a specialized MEMS chip that is designed to transform a compressive load of a picoindenter into a tensile load on the sample called Push-to-pull (PTP) device, was utilized and described in detail by Guo et al. [99].

2.5 Summary

Ns materials gain enhanced mechanical properties like high strength and hardness from the impediment of dislocation motion. In nc materials, this can be at the cost of a reduced ductility. Since TBs can act as sources but can also be penetrable, nt materials gain higher ductility. Changes in strength and ductility have also a strong impact on the fatigue behavior and the dominant mechanisms that are to be expected to be more GB related in the nc materials. In nt materials the fatigue is strongly affected by the dislocation-TB interaction and the detwinning processes. In the following four chapters, different aspects of cyclic and monotonic loading of ns fcc metals are presented. To include the influence of size and scaling effects, following tests were conducted:

- Fatigue of nc Ni microsamples

- Fatigue of nt Cu microsamples

- In situ TEM cyclic deformation of nc Cu nanosamples

- In situ TEM cyclic deformation of nt Cu nanosamples

The tests on microsamples which were in the range of a few mm were conducted on custom made and commercial tensile testing setups. The nanosamples were tested as ~ 50 - 100 nm thin films attached to a push-to-pull device *in situ* in the TEM.

3 Fatigue of nanocrystalline Ni

3.1 Abstract

The behavior of nanocrystalline (nc) Nickel under tension-tension fatigue was investigated. Commercial 220 µm thick nc-Ni foil (Integran) was purchased and cut by the use of a wire-EDM into hour glass shaped fatigue samples. The material had a grain size distribution between 20 and 80 nm. Those materials gained higher fatigue strength in the HCF regime compared to its coarse grained counterparts due to its higher tensile strength. SEM micrographs of cross sections near the crack tip revealed coarsening of the microstructure. Dimples in the fractured surface had a diameter of approximately 10 times the initial grain size.

3.2 Literature review

The microstructure of electrodeposited nanocrystalline Ni (ED nc Ni) has been described as being beneficial to mechanical properties, but being more prone to microstructural changes at the same time compared to its coarse grained counterparts [82, 11, 100]. Grain boundary (GB) mediated processes become much more likely as their volumetric content is increased. Examples are GB sliding and rotation where the deformation is accommodated by the GB without any influence of the grain interior [9, 30, 68, 69, 70]. Up to 50 % of the material volume is affected by the GBs, as grain size becomes less than just a few nm [7]. While the grain size is well below 100 nm and at a homologous temperature of T_m=0.17 the diffusion could play a significant role. At those small grain sizes, the diffusion path length is reduced significantly [101]. On the other hand, at the small grain size full dislocation mediated deformation is suppressed [67, 65]. Nevertheless, dislocation activity along GBs can enhance grain rotation and sliding [66]. For example the contribution of GB sliding to the overall deformation is estimated to be on the order of 20% in those materials [102]. Due to increased strength following the Hall-Petch relation [83] finer grained materials usually have also increased lifetime. The fatigue in nc metals and the dominant mechanisms are of current interest and are reviewed for example by Padilla et al. [85]. By reducing the grain size the materials experience a loss of ductility. Consequently, the deformation with a high plastic content leads to a low fatigue life in the low cycle fatigue (LCF) range (see Figure 3-1).

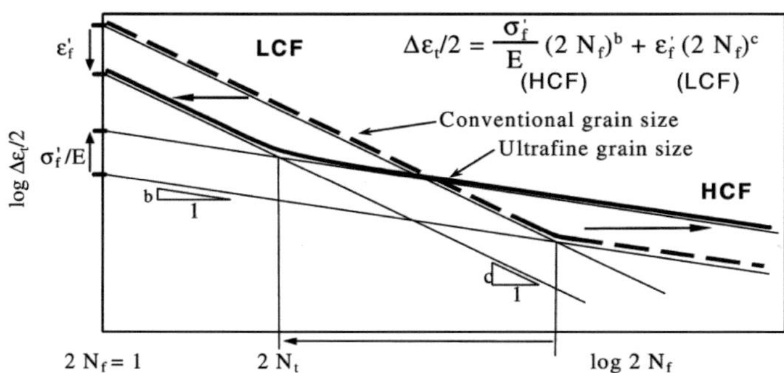

Figure 3-1: Schematic lifetime diagram (strain amplitude vs. cycles to failure) with LCF and HCF regime for conventional grained and ultrafine grained material (ufg) [103] [83].

As dislocation activity is constrained in these materials, failure is not taking place by conventional extrusion and crack formation [86]. Once a crack is formed, it can run very easily through the material as less crack deflection at GBs can occur [87] [88].

In conventional grained materials dislocation motion - and therefore material's deformation - during fatigue is mainly taking place in form of dislocations wall cell structures with a size of

$$d = \frac{KG|\vec{b}|}{\sigma} \tag{3-1}$$

[89] [104]. K is a dimensionless constant in the range of 10 [83] and the shear modulus G for Cu is 48 GPa. With a Burgers vector of $|\vec{b}|$ = 0.3 nm and an ultimate strength (UTS) in ultra fine grained (ufg) Cu of about 400 MPa this leads to a lower limit of the structure size of ~ 400 nm. Ultra fine grains with a size of 200-300 nm could not contain such structures as known from conventional materials.

For ufg Cu [90] as well as for nc Cu [91], it has been shown that grain coarsening of up to 30 % can occur during tensile-tensile fatigue. Similar results were found for nc Al tensile testing [105]. According to a representative failure distribution within the tested samples, the sample volume might also have some effect on the fatigue response of a sample. This was investigated by testing small samples [88, 93].

3.3 Experimental

The electrodeposited nc Ni foil (220 µm thick) used for the tests was purchased (Integran, Canada). For the tests, two different kinds of testing geometries were used: a dog bone shaped and an hour glass shaped geometry (Figure 3-2 a and b). The hour glass geometry was adapted with small modifications from [33]. The specimens have longer ear sections, in order to enable better gripping. For a good fit and alignment, the samples were mounted in dove tails that were machined with micro electro discharge machining (µEDM).

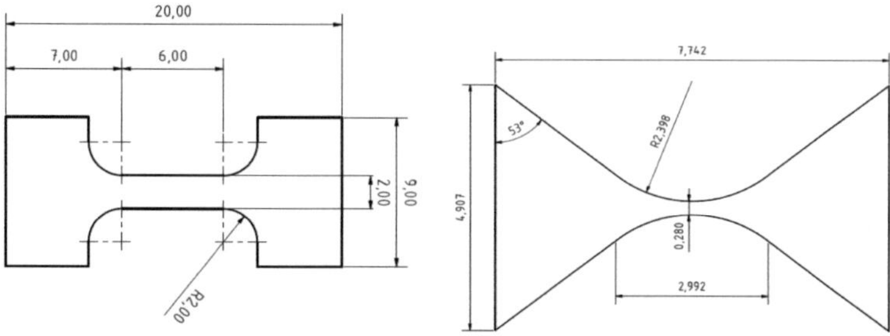

Figure 3-2: Technical drawing of the (a) dog bone shaped testing geometry and (b) the hour glass shaped geometry with longer ear sections for self alignment. The dimensions are in mm.

Some of the load controlled fatigue tests were conducted in an electro mechanic ElectroForce 3300 (Bose, Minnesota, USA) with the fatigue parameters listed in Table 3-1. The others were tested in a Eplexor 150 N (Gabo, Germany). They were conducted at 10 Hz and for the Eplexor with a static load of F_{stat}= 60 N (974 MPa) and a dynamic load of F_{dyn}= 32N (519 MPa). Examination of the fractured surfaces and the microstructure were done in Focused Ion Beam (FIB) Nova nanolab 200 dual beam system (FEI, USA). For the transmission electron microscope (TEM) sample preparation discs with a diameter of 3 mm were punched out mechanically from the 220 µm thick samples and then electropolished in a TENUPOL® (Struers, Germany). Best results were achieved with an electrolyte of 20% H_2SO_4 and 80% CH_3OH at a voltage of 15 V and a temperature of -10°C. TEM was carried out in a CM30 and Tecnai F20 at 200 kV (both Philips, Netherlands).

The grain size was analyzed using a custom written code for the line-intersection method implemented in Matlab® [106]. The size of the grains that are cut by the surface does not represent the statistic correct grain size distribution [107]. To correct that, a factor is used that includes the fact that grains are usually not cut at the largest diameter and are actually larger than they appear in the cross section. A factor of 0.79 for the correlation of the line length with the grain size was used.

Table 3-1: List of the used parameters for the fatigue tests in the ElectroForce 3300.

sample	min. load (N)	max. load (N)	load amplitude (N)	mean load (N)
1	10	170	160	90
2	10	120	110	65
3	10	90	80	50
4	10	110	100	60
5	10	110	100	60
6	10	110	100	60
7	10	110	100	60
8	10	110	100	60

3.4 Results

The fatigue test results, cycle numbers to failure at the various loading conditions are noted in Table 3-2. The dog bone-shaped specimens were used in these tests.

Table 3-2: Results of the fatigue tests of nc Ni. Only the specimen, marked with an x were fatigued until they fractured at the last measured cycle N_f.

Specimen	Broken	N_f ( )
1	x	7.5E+03
2	x	6.0E+04
3		1.0E+07
4	x	5.0E+06
5		2.5E+06
6		2.0E+06
7		3.0E+05
8		1.0E+05

The lifetimes of the fatigue tests are plotted in Figure 3-3. With cycle numbers between 10^4 and 10^7 cycles those experiments are in the HCF regime.

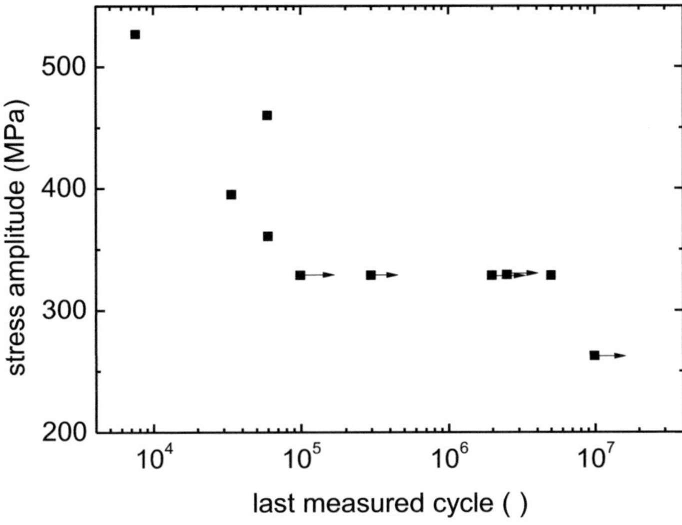

Figure 3-3: S-N (Wöhler) –plot of the tested nc Ni samples with a σ_m = 200 – 300 MPa.

No extrusions were found around the fractured surface of the fatigued samples (Figure 3-4). The fracture surface that failed perpendicular to the loading direction contains a region with a lot of striations. In the surface that fractured at 45° to the loading axis a lot of dimples evolved (Figure 3-4 d). The size of the dimples is on the order of ~ 200 nm, and therefore about 10 times larger than the initial grain size. The crack started in Figure 3-4 a at the left side of the sample at 45° to the loading axis. In the section normal to the loading axis the striations are curved against the crack growth direction.

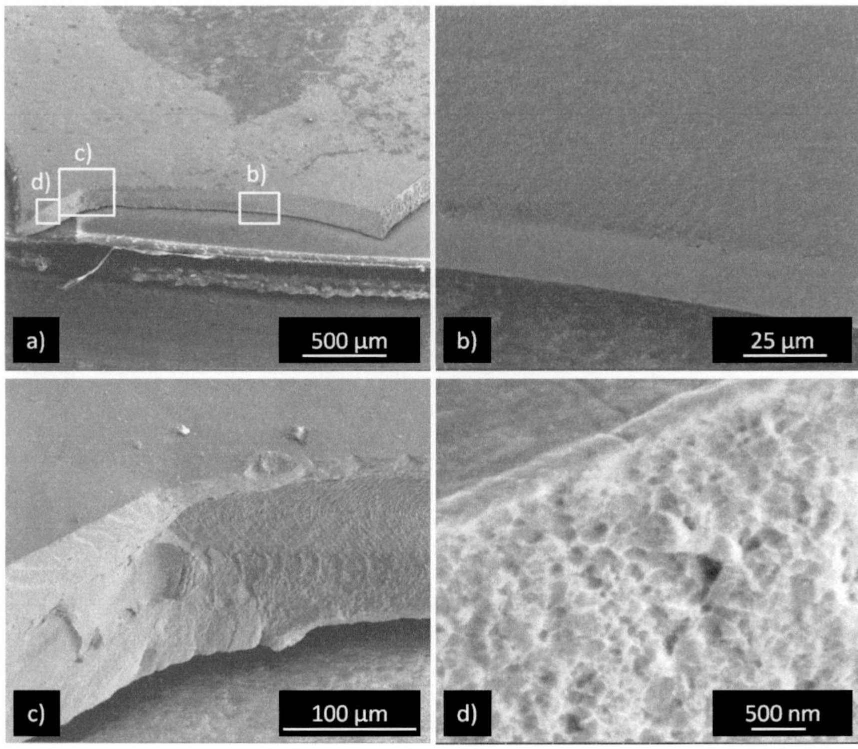

Figure 3-4: SEM micrographs of the fractured surface of a fatigued nc Ni specimen. (a) gives an overview over the fracture surface (b-d) with the detailed regions.

In Figure 3-5 a, the fractured surface of an hour glass shaped sample is depicted with an overview. It shows the different morphologies of the fracture surface in a region of ~ 30 µm along the side of the specimen compared to the inner parts. The cross section is shown in detail in Figure 3-5 b. It is visible that grains coarsened significantly about 1.5 µm beneath the fracture surface. The randomly oriented character of the surrounding grains of the initial state is transformed into elongated grains along the cross section. The coarsening in the different orientations is plotted in Figure 3-6 in a cumulative sum distribution vs. the grain size. The blue line represents the grain size distribution of the initial microstructure with a grain size between 5 and 100 nm. The red and black lines are indicating a coarsened microstructure along the flat side of the sample with grain sizes of up to 300 - 800 nm.

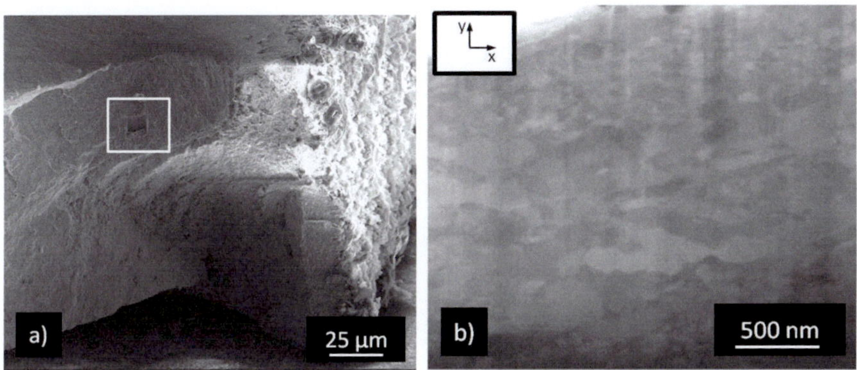

Figure 3-5: SEM micrographs of a fatigued nc-Ni hour-glass shaped sample of the (a) fractured surface and (b) the cross section. Some grains appear bigger than the average, which should be nc. The position of the cross section (b) is marked by the white rectangle in (a).

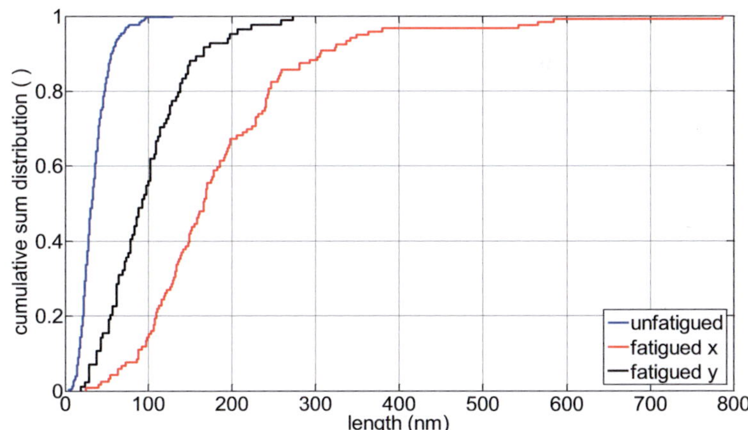

Figure 3-6: Cumulative sum distribution of the grain size for the coarsened nc Ni of Figure 3-5 b. The orientation perpendicular to the coarsened grains is labeled with "fatigued y" and along the coarsened grains with "fatigued x". The corresponding micrograph of the unfatigued sample is not shown here. The x and y direction are labeled in Figure 3-5 b.

Figure 3-7 a shows the as-deposited microstructure of nc Ni. In comparison the microstructure of fatigued nc Ni. Very few single dislocations were observed in bright field TEM (BF-TEM) in grains as small as 35 nm (Figure 3-7 b).

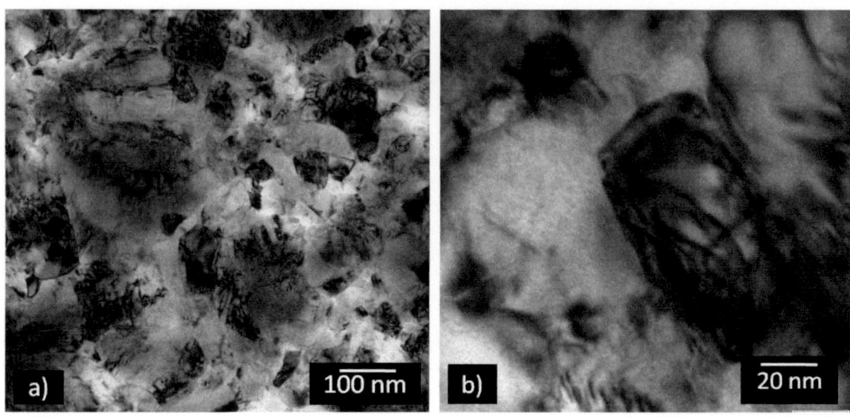

Figure 3-7: BF-TEM micrographs of (a) overview of as deposited nc Ni and in (b) fatigued nc Ni with dislocations within nanosized grains.

3.5 Discussion

In this section, the results from the last section are being discussed. At first, the fatigue strength of nc Ni and cg Ni are compared. In the following paragraph the influence of diffusion and GB mediated processes on grain coarsening will be estimated. Finally the extent of dislocation activity and the fracture behavior is critically discussed.

3.5.1 Fatigue behavior

The results showed an increase of HCF life of the nc materials in load controlled fatigue tests (SN-plots) compared to their cg counterparts (Figure 3-8). This is similar to what was proposed by Mughrabi for ufg materials [83] (see Figure 3-1). Even though, the fatigue strength of the nc material seems to be higher even in the regime at roughly 10^4 cycles and not just in the upper HCF regime. Stresses of roughly 300 MPa that induce plastic deformation in cg and probably also ufg material lead to a shorter fatigue life in ufg material due to less plastic deformability. In that case it makes sense, that the curves of the different grain sized materials cross in the total strain amplitude fatigue lifetime diagram. The situation has to be considered in another light for the nc material. The yield strength of the nc material is far beyond that of its cg counterpart. Stress levels of 800 MPa can be reached easily without any plastic deformation. Therefore, the detrimental conditions of plastic deformation are not reached for nc material and the fatigue life is higher than that of the cg material. The fatigue lifetime diagram in Figure 3-1 is not completely comparable to the S-N-plot of Figure 3-3 and Figure 3-8. Here, the strain-amplitude is plotted versus the cycles to failure, as the displacement control of the total strain amplitude versus lifetime plot leads to another behavior of the material compared to the stress-control of the S-N-plot. In the presented experiments a tensile mean stress was applied to prevent buckling of the samples. This resulting in a positive stress ratio of $R = 0.1$. In Figure 3-8 the lifetime values of the samples measured here are plotted in comparison to fatigue lifetimes for notched and unnotched cg Ni from [108].

Similar behavior was seen for microcrystalline (mc) and nc Co [109]. Down to $2 * 10^3$ cycles to failure in a load controlled fatigue test, the curve of mc Co is below that of nc Co in a Wöhler (S-N) plot. It seems likely, that the deformation stays elastic in the nc material at this fatigue lives while it is plastic in mc Co. Nevertheless the applied maximal stresses are by a factor of two higher for nc material, which is in good agreement with observations in literature for Ni. The correlation between the longer fatigue life of the nc material and the time needed for grain coarsening to a stage, where conventional fatigue processes can act stays unclear, especially in the elastic HCF regime. These processes include dislocation motion which is expected to dominate in the nc regime between 50 and 100 nm [82]. Full dislocation networks and extrusion formation starts in the regime of grain sizes of a few hundred nm and more [110]. The grains in Figure 3-5 b grew to a size, where PSB formation is expected to appear. Even though, no extrusion formation was observed. A reason might be, that the coarsening did not appear in the region directly underneath the surface and nc material between the large grains and the surface prevented the PSBs to form.

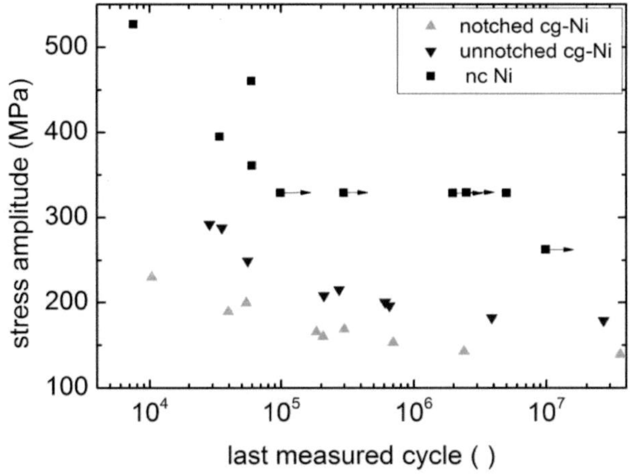

Figure 3-8: Wöhler (S-N) -plot: the cg Ni values are from [108] where a σ_m (cg-Ni) = 200 MPa and in this work a σ_m (nc-Ni) = 200 – 300 MPa was used.

3.5.2 Coarsening by GB diffusive and rotational processes

The coarsened structure in the FIB cross section Figure 3-5 b does not seem to be a result of ion radiation as this leads to a more isotropic structure [111]. Furthermore, thermal coarsening can be excluded, as this would apply to the whole structure and would not just appear locally. It is indicated in the cumulative sum distribution of Figure 3-6 that not just the larger grains coarsened during fatigue. Especially in the x-orientation one grain coarsened to a length of more than 600 nm in band like structures. In the perpendicular y orientation the biggest grain was on the order of 300 nm.

Several effects are expected to play a role during coarsening of nc metals. The small grain size implies that a big amount of GBs are affected by diffusion processes. Therefore, diffusion along GBs is discussed to have a big effect on the coarsening and the deformation mechanisms [101]. Due to the limited availability of diffusion data of nc materials and the difficult determination of diffusion constants, values for the material used in this work are extrapolated from literature values. In [112] the GB diffusivity of Cu in nc ED Ni was measured. The diffusivity of Cu and Ni should be very similar due to the almost equal atomic weight and the equal structure and melting temperature. It is clear, that the values are not the same and this was only done to estimate at least the order of magnitude of the diffusive behavior.

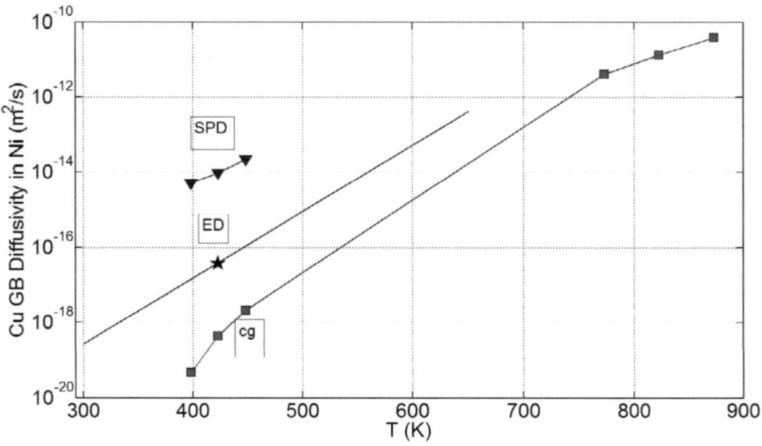

Figure 3-9: GB diffusivity of Cu atoms in cg Ni, as well as SPD and ED Ni (from [112]))

A diffusivity for GB diffusion of Ni in nc Ni at RT was estimated to be between 10^{-20} and 10^{-19} m^2/s. This was done by extrapolating the value for ED Ni to a RT with the slope of the cg and SPD Ni in Figure 3-9. It is not desirable to plot a trend line for the estimation, containing only one data point. As the slopes in the diffusivity vs. temperature diagrams are expected not to be strongly dependent of the grain size and this slope is only used to estimate a range of diffusivity at RT, these assumptions are considered to be valid. The GB diffusivity from literature at 293 K of Cu in nc Cu is on the order of D_{GB} = 2.6 * 10^{-20} m^2/s [113, 114]. This value is roughly in the range of the estimated value. After geometrical considerations in a cubic lattice, a grain rotation of maximal 45° is needed to rotate one grain orientation into the orientation of a neighboring grain. At a grain size of roughly 10 nm, this correlates to a diffusion path length of 4 nm along the GB. With Eq. (2-7) a time between 27 s and 267 s can be estimated, when a D_{GB} of 10^{-19} and 10^{-20} m^2/s is used. A combination of diffusion and glide of dislocations could lead to grain rotation of neighboring grains into one preferred orientation and as a result to coalescence. This sliding and diffusive flow was already discussed by Ly et al. [34]. It might be an important process in nc materials as it enables the coarsening and coalescence at room temperature to an extent, that is known from cg materials to only happen at elevated temperatures. The source that enables these processes is the small grain size and the short diffusive paths instead of energetically unstable GB structure like in SPD materials. Impediment of coarsening could arise from interlocking grains which lead to steric hindrance and prevent rotational deformation. But this effect is expected to be fairly small in our experiments as the grains that are used in this consideration adopt a fairly round geometry, as this provides the best ratio of volume to surface energy.

The rotation process by dislocation motion was recently presented by Bobylev et al. [115]. Shear deformation leads to a pile up of lattice dislocation at triple junctions. It can transform into grain rotation, assisted by diffusion assisted climb of GB dislocations. To achieve coarsening of the microstructure to sizes of several hundreds of nm as visible in Figure 3-5 b, the above mentioned process has to be applied in a repeated manner in order to adapt adjacent grain orientations. Each grain has only a diameter of maximal 100 nm. If only the stress concentrators from lattice dislocation pile ups are considered, it could take up to 2670 s if a region of 10 grain diameters is rotated into the same orientation layer by layer in both directions.

3.5.3 Dislocation activity

Single dislocations were observed by TEM in nc Ni down to grain sizes of ~ 30 nm, even though they were very rare. By using Eq. (3-1), the minimum grain size where grains could bear full dislocation structures can be estimated to be on the order of 80 nm, by using a K of 10, a shear modulus of $G = 48$ GPa, a Burgers vector of $|\vec{b}| = 0.3$ nm and a strength of $\sigma_{uts} = 2$ GPa. For smaller grains it would cost too much energy to force dislocations into such small grains. Therefore, the formation of classical dislocation structure and the corresponding dislocation based coarsening mechanisms can be excluded for metals with the here found nc microstructure.

Due to the low dislocation activity, extrusion formation is significantly reduced in grains that are smaller than 1 µm [110], but were still observed on the surface of fatigued ufg Cu [90]. Persistent slip bands which are a precursor for extrusions [84] were observed in fatigued nc Ni that coarsened to a ufg material [116]. Neither extrusions nor persistent slip bands were found in the fatigued nc Ni samples in this work. In the initial state, the grain size does not allow dislocation structures to develop, which are needed to form persistent slip bands and then extrusions [110]. It could not be explained why dislocation structures did not appear in the coarsened microstructure of this work. One reason might be the load control of the fatigue experiments. Once the microstructure starts to coarsen and loose some of its strength provided by the Hall-Petch strengthening, loading might get more detrimental. The stress levels, that were in a moderate regime before might lead to fatal failure in the coarsened and weaker material. Consequently, the microstructural stability is an important factor for the fatigue and cycling properties of nanostructured metals.

3.5.4 Fracture behavior

The coarsened grains (Figure 3-5 b) are roughly 1.5 µm away from the fractured surface. This means these grains are not directly at the crack tip but could be influenced by the plastic zone. With

$$r_p \sim \frac{1}{2\pi}\left(\frac{K_Q}{\sigma_y}\right)^2 \qquad (3\text{-}2)$$

from [74] the size of the plastic zone can be estimated to be on the order of ~ 7 µm. A fracture toughness on the order of K_Q ~ 10 $MPa\sqrt{m}$ [117] and a yield strength of σ_y ~ 1500 MPa was used. Though the sample is big enough to bear the plastic zone. The coarsened grains are expected to be in the mechanically influenced plastic zone.

Over all, the fracture surface is more even in the flat specimen with the dog bone shape than in the hour-glass shaped geometry where it is more faceted (see Figure 3-4 & Figure 3-5 a). The different fracture morphology in the two different sample geometries is due to different aspect ratios of the sample thickness to width. The transition in terms of the roughness of the fracture surface was related to a transition from the intermediate to the high ΔK regime [116]. These structures in the middle part which are magnified in the upper right image were already observed and discussed [37]. Questionable remains if the grains coarsened to an extent, where the dimples are in this range or if the deformation mechanisms that formed the dimples work on a length scale which is roughly ten times the initial grain size, what was already observed before [82]. They only appeared in the fracture surface which is under 45° to the loading axis. This part is expected to be the site where the crack was formed and grew steadily, before fatal rupture occurred in the part that is aligned perpendicular to the loading axis.

3.6 Summary

The main implications of fatigue in nc Ni that were elaborated in this work are:

1) HCF fatigue lifetime and also LCF is strongly increased compared to cg materials in a load controlled test.
2) Conventional dislocation motion is not expected to lead to PSB's and extrusions and finally failure, as the grains are too small to bear full dislocations networks.
3) Other mechanisms than the conventional dislocation motion could be the cause of grain coarsening to the extent observed here. The coarsened grains then might have a reduced strength and therefore will be prone to carry the plastic deformation leading to failure.

3.7 Acknowledgement

Acknowledged is the help of Dr. Vitor Barroso, Dipl.-Chem. Johannes Höpfner for their help with the Gabo Eplexor and Prof. Hael Mughrabi for fruitful discussion.

4 *In situ* TEM mechanical testing of nanocrystalline Cu

4.1 Abstract

Nanocrystalline (nc) metals have promising mechanical properties due to increased interaction of grain boundaries and dislocations. In exploring the prevalent mechanisms that govern the deformation, it is more valuable to see what happens during the test instead of just seeing the initial state and the state after the deformation. Therefore, *in situ* TEM tensile tests were conducted in a displacement controlled manner on nc Cu thin films with a thickness of ~ 62 nm and grain size between 10 and 100 nm. Especially for nc metals it is still unclear what happens during the loading in single grains. If dislocation activity plays a significant role in the deformation behavior or if the main mechanisms are grain boundary (GB) related. On a larger scale, the microstructural evolution and cracking is observed, to see how the grains and the GB structure develop. The high GB content causes a high ultimate strength of ~ 2.5 GPa. The Young's modulus is of the order of 70 GPa and by a factor of ~ 2 lower as expected for mainly (111) oriented Cu. During cyclic loading grain coarsening took place, what further affects the dominating deformation mechanism.

4.2 Theory

Nc materials have improved strength and hardness compared to their coarse grained counterparts [32, 118] and therefore are attractive materials for many applications (e.g. hard coatings, integrated micro systems...). These enhanced mechanical properties are gained from the high GB density and increased dislocation constrains following the Hall-Petch relation. On the other side, this also causes the microstructure to become less ductile and thermally unstable.

Since the nanosized grains are too small to bear any remarkable dislocation networks a transition takes place [119] where more GB related mechanisms are dominant. Examples for possible mechanism are GB diffusion [120, 85] and diffusionless stress induced GB motion [30]. Grain rotation [69, 121, 70] and/or GB sliding [122, 123, 124] could become the governing deformation processes by dislocation climb. Simulations showed that partials could be emitted from a GB and move through a grain [125, 126]. Trailing partials on the same glide plane can remove the stacking fault of the leading partial, which leads to a full dislocation of the crystal. If the trailing partial moves on a parallel glide plane, a twin is the result. Since the mechanical properties are strongly tied to the GB structure and grain size, its evolution is of great interest.

Another parameter that will be discussed is the Young's modulus. As the conditions of the tested samples do not meet the conditions for the classical understanding of the Young's modulus (for example some porosity) it will be described as the stiffness in the following text. For not textured coarse grained (cg) Cu the Young's modulus is 128.0 GPa. Sputtered or electrodeposited films usually develop some texture, predominantly <111> out of plane, with a modulus of 191.1 GPa. The <100> orientation is more pronounced for very thin films (e.g. a few nm) and has a value of 66.6 GPa. The <110> orientation has a value of 130.3 GPa, although it can be rarely found in these films [127, 128].

Nc materials are prone to coarsen during cyclical mechanical loading and this can reduce the mechanical performance [90, 91]. In other cases hardening was observed during fatigue by depletion of dislocation sources [92]. This change could also take place in the nanostructured metals investigated in this study.

GB related processes seem to dominate in the nc regime at the expense of dislocation based plasticity [85], what is the dominant deformation mechanism in cg Cu. As films get thinner, extrusions lose importance due to dislocation constraints. Therefore, extrusions are smaller at grain sizes smaller than 1 μm [110]. To investigate the underlying mechanisms, mechanical tests under controlled mechanical conditions are necessary.

Due to the sought small size of the samples which is necessary for *in situ* TEM testing, the samples have to be prepared in a laborious technique where focused ion beam (FIB) is involved. As a consequence the FIB-damage and their effects on the micromechanical test have to be taken into account [129]. Possible consequences are implantation of Ga^+-ions in the material. This could lead to embrittlement, as the Ga^+ can segregate to the grain boundaries and reduce the GB strength leading to GB decohesion [130, 131].

For the determination of the fracture toughness K_{IC} standardized equations for ISO samples can't be used, as no specific sample geometries were used. Instead the following equation after [74] is used for plain strain conditions

$$K_{IC} = \sqrt{G_C * E} \tag{4-1}$$

E is the Young's modulus or stiffness and G_C is the critical energy release rate. This is the difference of total introduced energy dW and elastically stored energy, so the plastically stored energy dE available for crack growth, divided by the area created by the growing crack.

Nc materials are often available in the form of thin films due to the common methods of deposition, like sputtering or galvanoforming. Accordingly, the set-ups used for testing have to be able to provide the necessary accuracy. The strength of nc Cu microsamples was indicated by tensile tests on nc Cu thin films [132] and bulk samples [133]. The idea of using MEMS (micro electric-mechanical system) to apply load on thin films was presented by Haque and Saif [94, 96] as well as by Gianola and Eberl [134].

The aim of this work is to conduct displacement controlled *in situ* tensile and cycling tests on nc Cu in the TEM to actually observe the deformation during cyclic loading.

In the following, the sample deposition and preparation are explained and the results for tensile and fatigue tests in terms of evolution of the microstructure and mechanical characteristic are presented.

4.3   Experimental

4.3.1   *In situ* setup

The condition for *in situ* TEM testing samples is the electron transparency which means a thickness of less than ~ 100 nm. To be able to apply a defined load and displacement to those small samples the concept of a MEMS chip was chosen. The testing setup that contains a picoindenter was already used for *in situ* compression tests [135, 136]. For the *in situ* tensile loading a so called push to pull device (PTP) was used [99]. This PTP converts the compressive load of the picoindenter that is pushing on the bump of the PTP into a tensile load on the sample that is sitting on the PTP (see Figure 4-1). At the beginning, the Cu films that are attached to the PTP (PTP device, Figure 4-2) have to be cut in the right geometry in the FIB (FEI Strata 235 Dual Beam FIB, Netherlands) between the crossheads (at 100 pA / 30 kV) (see Figure 4-1 right). Also the film between the gaps of the PTP aside the sample has to be removed at 1nA / 30 kV (Figure 4-1 left).

The displacement controlled *in situ* TEM tensile tests were then conducted in a TEM (JEOL JEM 3010, Japan). The vertically loaded sample can be observed during deformation on the PTP through the little gap from the top.

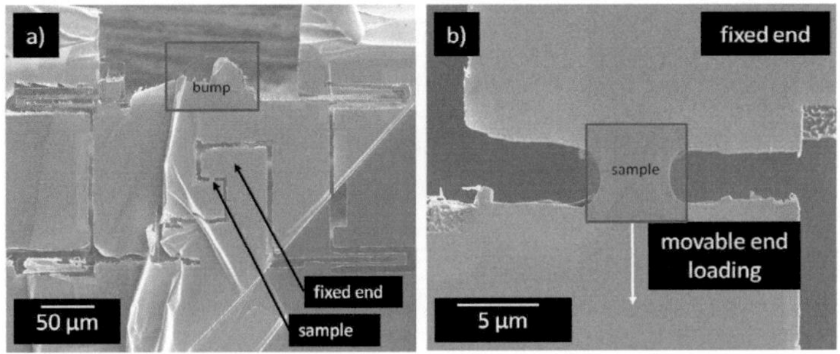

Figure 4-1: SEM-micrograph of thin film sample machined in the FIB and attached to a PTP. a) thin film on the PTP b) shaped tensile testing sample.

4.3.2 Thin film deposition

The Cu specimens were deposited by radio frequency (RF) magnetron sputtering using a 2″ diameter planar target with 99.99 % purity and were provided by a group at the Institute of Nanotechnology at the KIT. The Cu target is facing the rotating sample stage at a distance of 150 mm, at an angle of 20°. The base pressure of the sputtering chamber was $2.0 \cdot 10^{-8}$ mbar. At room temperature the Cu films were deposited in 5 steps of 10 nm (= 50 nm) thickness. Between these steps, the deposition was interrupted for 10 s using a fast rotational shutter in front of the Cu-target, to suppress columnar grain growth [137, 105]. Sputtering was performed at 0.005 mbar Ar working gas pressure with 100 W sputtering power. The substrates in the shape of 10 x 10 mm mica plates were cleaved into thin foils and coated first with 5 nm carbon using a Carbon Coater (Cressington, England). Due to the low thermal stability of nc Cu, the samples were stored in a refrigerator at a temperature of ~ - 18 °C. The XRD measurements for calibration of the sputter process and the texture analysis were conducted by a X-ray diffractometer (Bruker AXS D8, USA) and Cu source (Cu K_α with 1,544 Å).

The thickness of the deposited films was measured again by FIB and atomic force microscopy (AFM) (MultiMode Nanoscope III Veeco, USA). Before the SEM measurement a Pt bar was deposition at the cross section and the micrograph was captured in immersion-mode to get a better resolution. For the AFM measurement, a small piece of the film was removed from the mica and attached to a glass plate. Line profiles across the edge were acquired and step heights of the film were determined over 5 profiles.

4.3.3 Sample preparation

To be able to test the thin films with PTP devices the thin films had to be transferred first. Therefore, the mica substrate was dissolved in distilled water. In the liquid the Cu peels off and floats at the surface. In the next step the piece of film has to be caught with the PTP device (see Figure 4-2) carried by tweezers.

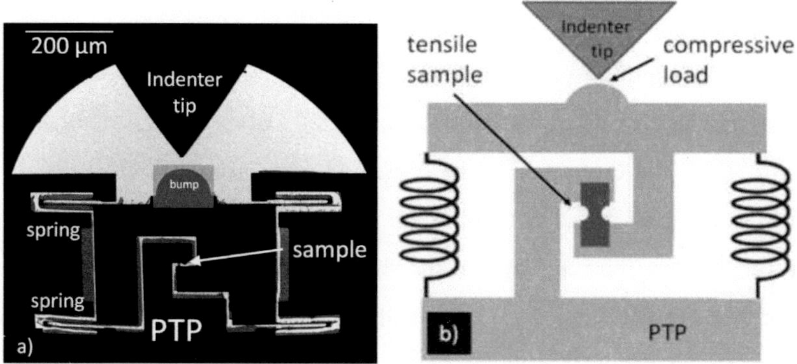

Figure 4-2: (a) BF-TEM of the PTP with the sample in the testing region and the tip of the picoindenter. (b) schematic illustration of the PTP configuration .

To remove the Cu from the PTP after testing, they were treated in acid (aqua regia). Therefore the used PTP device with the 60 nm Cu-film is placed in a solution of hydrochloric acid (35 %) and nitric acid (65 %) in a ratio of 3:1 for about 30 min. It is important to flush it with sufficient distilled water afterwards to keep the gaps of the PTP clear. Alternatively, the Cu can be milled away with the FIB.

4.3.4   Test procedure

In the next step, the indenter tip was approached to the bump of the PTP up to a distance of 50 µm. During the approach of the tip it is safer, to focus on the sample, as contrast changes in the grains indicate even the slightest contact, even though the load was not increasing (less then µN). Once the tip was in contact, the indenter was moved back for maybe 50 nm to choose the loading parameters like maximum displacement and displacement rate.

For the tensile test, a displacement-time segment of 800 nm in 80 s was chosen. Videos were taken during the tests in order to calculate the strains by digital image correlation (DIC) in a post process. This matlab® code uses fiducial markers that stick to surface structures with a high contrast difference and follow the deformation in order to digitize it [138]. One difficulty with the application of DIC on TEM micrographs is that changes in contrast are quite often. This is a big issue for the DIC-algorithm as it orientates on contrast edges in the single images. The contrast changes can happen due to bending contours, or just when any other mechanism changes the diffraction conditions of the observed grains/area.

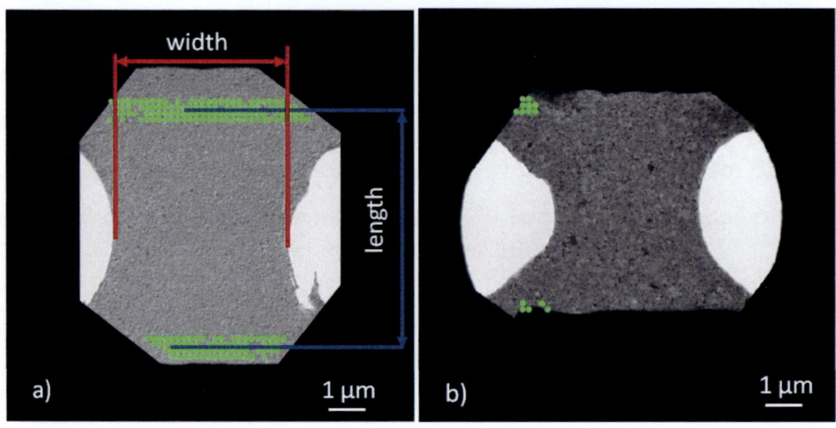

Figure 4-3: Positions of the grid-markers for the DIC are green. The strain is measured in regions, where almost no strain and therefore no contrast changes occurred. (a) sample ncCu62nmT1 and (b) sample ncCu62nmT3.

The displacement amplitude for the cycling test was set between 50 and 120 nm. After each cycle the load at zero displacement was different, due to plastic deformation and had to be readjusted to a value between 10 and 15 µN (σ_U) to prevent compression.

The error of the strength of the samples from this work was calculated by a Gaussian error propagation with a load F and deviation ΔF of 8 µN in load, of thickness y and thickness deviation Δy of 5 nm and in width x and deviation in width Δx of 20 nm.

$$\Delta \sigma = \sqrt{\left(\frac{1}{xy}\Delta F\right)^2 + \left(-\frac{1}{x^2}\frac{F}{y}\Delta x\right)^2 + \left(-\frac{1}{y^2}\frac{F}{x}\Delta y\right)^2} \qquad (4\text{-}2)$$

The cumulative grain size distributions were acquired in a custom made line intersection function in matlab ® [106].

The following labeling convention is used: The name of each sample consists of the material (nc Cu), the film thickness (62nm), the testing mode (T for tensile, C for cyclic testing) and an ongoing number.

4.4 Results

4.4.1 Sample size and defect density

According to the chosen sputter time, the thickness should have been 50 nm. In the SEM a film thickness of 65 nm was measured in a cross section with the nc Cu between the mica substrate and a Pt bar (Figure 4-4 a). In the AFM the thickness was around 59 nm on a piece of film attached to a glass plate (Figure 4-4 b). For the following calculations a mean value of 62 nm was used. In the same AFM -mapping a film roughness of 1.73 nm (RMS) was determined.

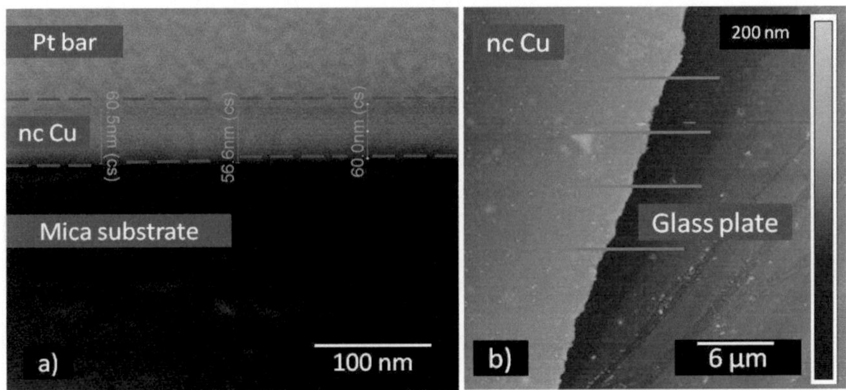

Figure 4-4: Thickness measurement of the nc Cu films: (a) SEM-micrograph of a cross section of nc Cu. (b) AFM mapping of the edge of a nc Cu film attached to a glass plate. The blue horizontal lines indicate the profiles where the height was measured.

4.4.2 Microstructure

The XRD spectra of the investigated nc Cu thin films are plotted together with a generated spectrum of an isotropic powder in Figure 4-5. At 43.4° the dominant [111] peak is visible for nc Cu thin films, and at 50.4° the [200] peak. The [100] peak is erased in face-centered cubic metals like Cu. In the isotropic case of a powder, the ratio of the [111] and the [200] peak is I_{111}/I_{200} ~ 0.47. In our measured curve, the ratio is I_{111}/I_{200}~ 0.30, which reflects a microstructure with a preferred orientation in [111], but also approximately one third of the grains are oriented in [200]. The signal to noise ratio (SNR) is fairly small due to the thin thickness of the sample (~ 62 nm).

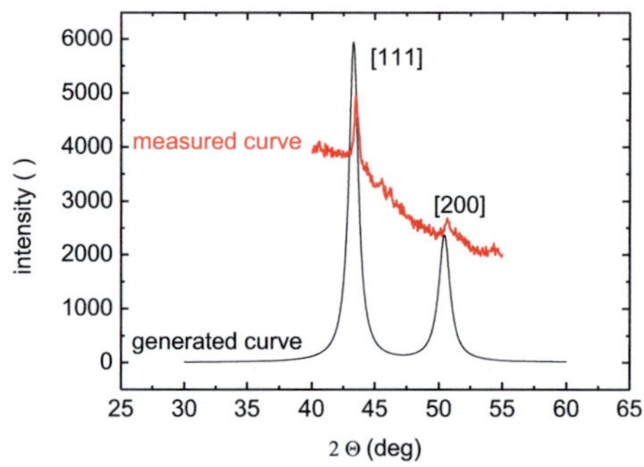

Figure 4-5: Theoretical XRD spectra, generated with powercell ® of Cu and the measured curve of nc Cu

The initial microstructure is depicted in Figure 4-6 a. The mean grain size is right below 20 nm and the largest grains are up to 60 nm in diameter. In HRTEM some microvoids in the range of up to 8 nm in diameter were found (see Figure 4-6 b). The distribution is not quantified, but the round pores appear over the whole sample at triple points. During the tensile test, no obvious coarsening of the polycrystalline microstructure was observed.

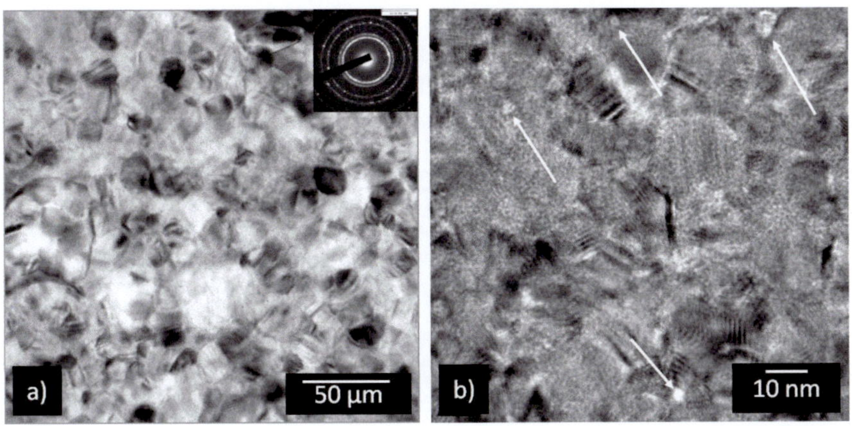

Figure 4-6: TEM micrograph of a as deposited nc-Cu sample, (a) the microstructure and (b) HRTEM image of nc Cu microstructure. Microvoids with a size of a few nanometers can be observed at the locations marked by the white arrows.

The cumulative grain size distribution of the samples ncCu62nmT1 and ncCu62nmT3 are depicted in Figure 4-7. The microstructure of ncCu62nmT1 is completely in the nc regime, whereas ncCu62nmT3 has some ufg components.

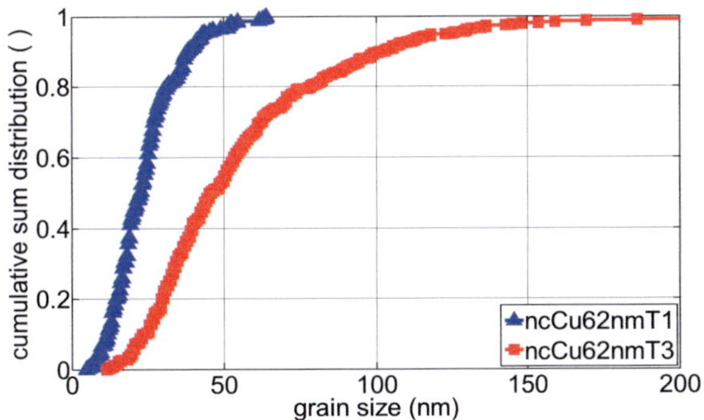

Figure 4-7: Cumulative grain size distribution of micrographs of ncCu62nmT1 and T3 which are depicted in Figure 4-3.

4.4.3 *In situ* tensile test

Table 4-1 lists the results for the ultimate tensile strength (UTS) from the tensile tests on ncCu62nmT1 and ncCu62nmT3 and the last cycle of the cycling tests. The results from the tensile tests are depicted in Figure 4-8.

Table 4-1: List of tested samples in the tensile and cycling tests at used loading rate and resulting UTS. The loading rates are documented for the last cycle. The nominal strain rate was calculated for a gage length of 2200 nm.

sample	cycles	disp. rate (nm/s)	$\dot{\varepsilon}$ (1/s)	UTS (MPa)	σ_y (MPa)
ntCu62nmT1	1/4	8	$3.6*10^{-3}$	1653 +/- 167	980
ntCu62nmC2	39	30	$1.4*10^{-2}$	523 +/- 69	400
ntCu62nmT3	1/4	10	$4.5*10^{-3}$	2449 +/- 242	1500
ntCu62nmC4	23	9	$4.1*10^{-3}$	1326 +/- 134	840
ntCu62nmC5	47	140	$6.4*10^{-2}$	1656 +/- 168	1110

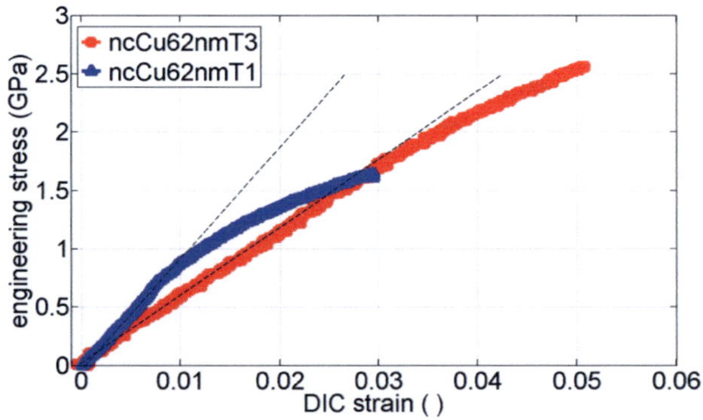

Figure 4-8: Engineering stress vs. DIC strain, measured in the region that is marked in Figure 4-3 a and b at a thickness of 62. For both tests a PTP with a stiffness of around 20 N/m was used. The slope of the sample stiffness is indicated by the dashed lines.

The tested film had a maximal ultimate strength of roughly 2.5 GPa and 1.6 GPa and a total DIC strain of 5 % and 3 % (Figure 4-8). In an estimation of the local deformation in the narrowed middle part, a 50 % higher strain can be expected due to the shorter referential length (see Figure 4-3).

In the two tensile tests, a stiffness of 94 GPa (ncCu62nmT1) and 55 GPa (ncCu62nmT3) was measured at the slope of the elastic segment (Figure 4-8). The fracture appears to be brittle as is visible in Figure 4-8. Also the fracture surface is straight without any noteworthy necking in thickness. This goes in hand with the stress-strain-data which indicates almost no plastic deformation in ntCu62nmT3. This sample also failed brittle but showed more strain than the ntCu62nmT1 sample. In that sample a crack formed at the end of the last loading segment (Figure 4-9).

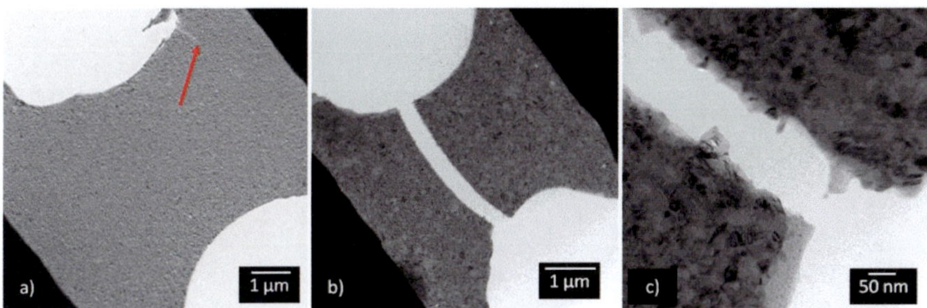

Figure 4-9: BF-TEM micrographs of (a) ncCu62nmT1 where a crack formed. The crack tip is indicated by the arrow. (b) ntCu62nmT3 failed just brittle. (c) close view of the fractured surface which appeared to be intercrystalline with some amorphous layer at the edge of the sample.

4.4.4 *In situ* cycling test

Grain coarsening was visible in several samples for example in ncCu62nmC4 (Figure 4-10 a and b). The sum distribution of different stages of fatigue show a trend that the grain size distribution shifted to higher values. Especially the fraction of grains in the ufg regime increases (Figure 4-11).

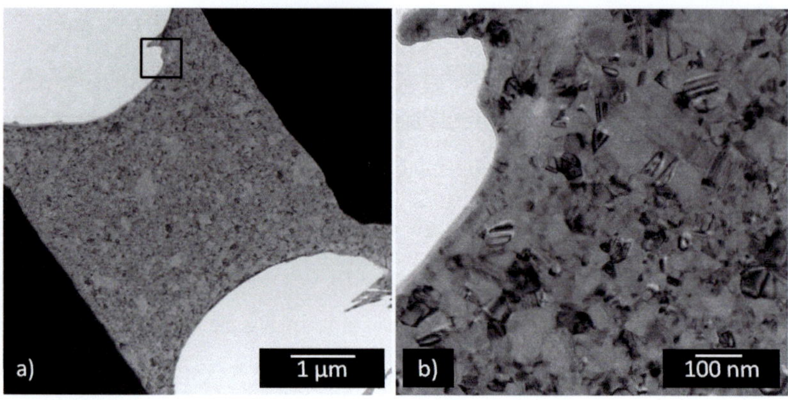

Figure 4-10: BF-TEM micrographs of the cycled sample ncCu62nmC4. The rectangle in (a) indicates the region where (b) was taken

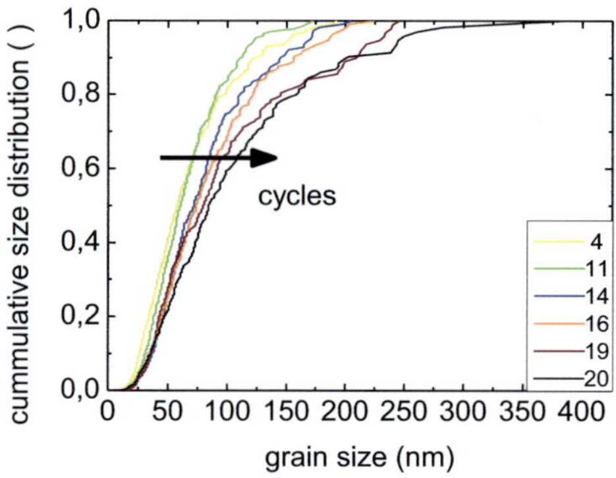

Figure 4-11: Cumulative sum distribution of the grain size distributions after 4, 11, 16, 18, 19 and 20 cycles of the microstructural development of ncCu62nmC4 in the region shown in Figure 4-10 b.

The development of the microstructure in sample ncCu62nmC5 during the 9th cycle around a FIB notch is depicted as a series of BF-TEM micrographs during *in situ* cyclic loading (see Figure 4-13). The TEM micrographs were taken at the notch at several load levels which are marked in Figure 4-12. The notch opened between a and e while the crack formed in f. It grew until g and closed in picture h and i.

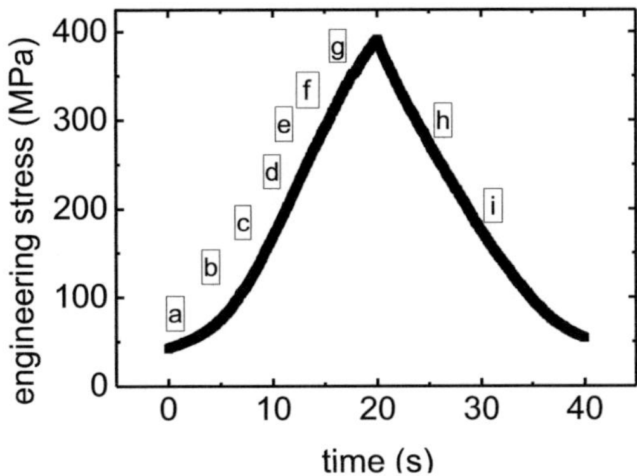

Figure 4-12: Engineering stress vs. time of the 9th loading cycle of ncCu62nmC5. The letters mark the load levels, where the micrographs from Figure 4-13 were taken.

Initially, the crack started in the notch which is coated as the rest of the sample with a ~ 20 nm thick amorphous layer. The crack propagated along the GB of the center grain and stopped there after maybe 30 nm. The crack opened and closed during loading and unloading, clearly visible by the contrast change when the crack flanks came in contact during unloading. The other part remained in some kind of surface layer which contained oxide and remaining carbon from the deposition process that went not completely off when the films was floating off the mica plate. Between Figure 4-13 e and f the step on the right side of the big central grain vanished due to the movement of the upper part of the GB (indicated by the white arrows in both micrographs). At the same time, during loading between e and g, it appears that the big middle grain grew under the influence of the crack in a shift of the right GB to the right. Severe strain is accommodated in the upper part of this grain (Figure 4-14).

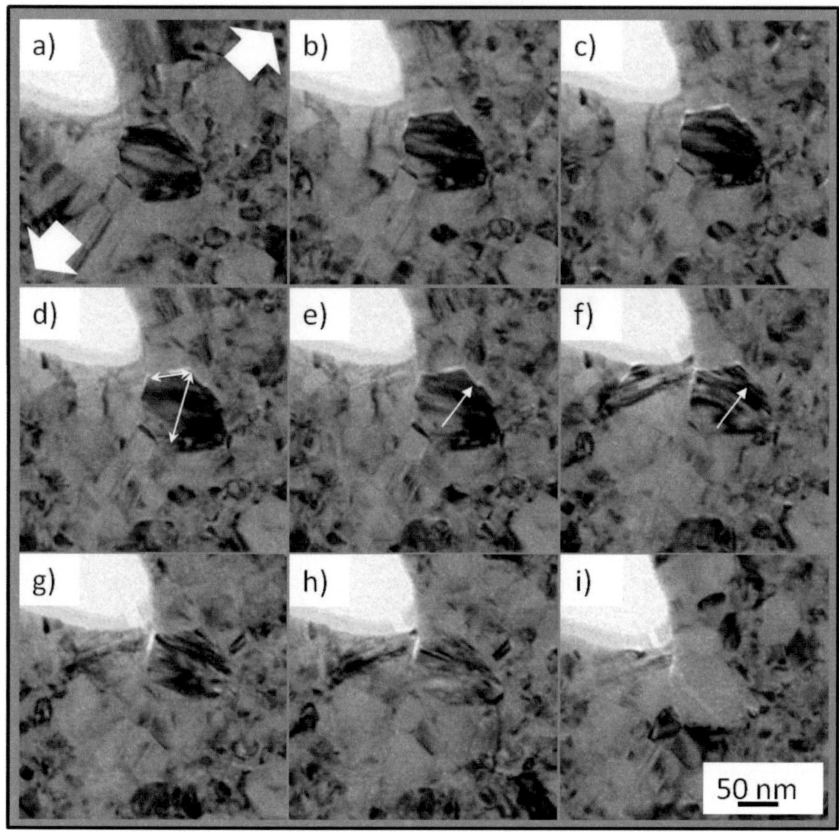

Figure 4-13: Micrographs of the *in situ* test on ncCu62nmC6 during 9th cycle (crack initiation). The sample was loaded until cracking in picture (f) and further to (g) and then unloaded in picture (h) and (i), where the sample is completely unloaded. The loading direction is marked with the white arrows in picture (a). The arrows in micrograph d show the position where length and the diagonal of the central grain were measured.

Figure 4-14 visualizes that both, the length of the upper edge of that grain and the grain diagonal, increased during loading. Some plastic deformation appeared in the upper part of the grain, as the length stays constant after the loading. After the 9th cycle, the crack did not significantly grow but was opening and closing with loading and unloading.

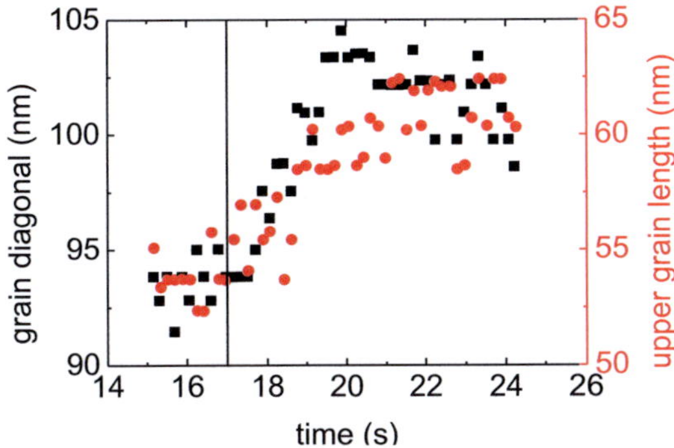

Figure 4-14: Plot of the length grain diagonal (left axis) and the length of the upper edge of the center grain (right axis) vs. the duration of that cycle between 45 and 85 s of the test. The vertical line in both images marks the image, where the crack was formed. The position of the measured lengths are shown in Figure 4-13 d.

In sample ncCu62nmC5 the displacement rate was increased from 130 nm to 1400 nm in 10 s for the 47th cycle thereby a crack initiated at the notch tip. The crack extended into a nanotwinned grain right at the bottom of the notch tip. In Figure 4-15 the different stages of the growing crack are depicted. In micrograph a, the sample is in the initial state of the 47th cycle. The nanotwinned grain is clearly visible, with several twins labeled with T1-T4. The loading axis is implied by the white arrow. In figure b the crack has grown into the amorphous layer at the edge of the notch. Similar to the cracking in Figure 4-13, the carbon layer cracked. By now, the crack has grown into the nt grain about 20 nm beside the GB and T1 fractured in c. The tip of the crack is shown clearly with the white arrow. In pictures d and e the crack extended and T2 and T3 fractured stepwise intercrystalline and not along the GB which is roughly 20 nm away and marked with "GB" (Figure 4-15 f). Interestingly the crack progressed beyond this grain, as the last twin T4 was not completely fractured in picture f what is an indication for crack bridging. As some shear load seemed to be applied during loading, the fracture mode was forced into some side movement. Hence the grain seems to have rotated. Final fracture occurred right after the last image. Until that stage, the crack has grown to a total length of roughly 100 nm.

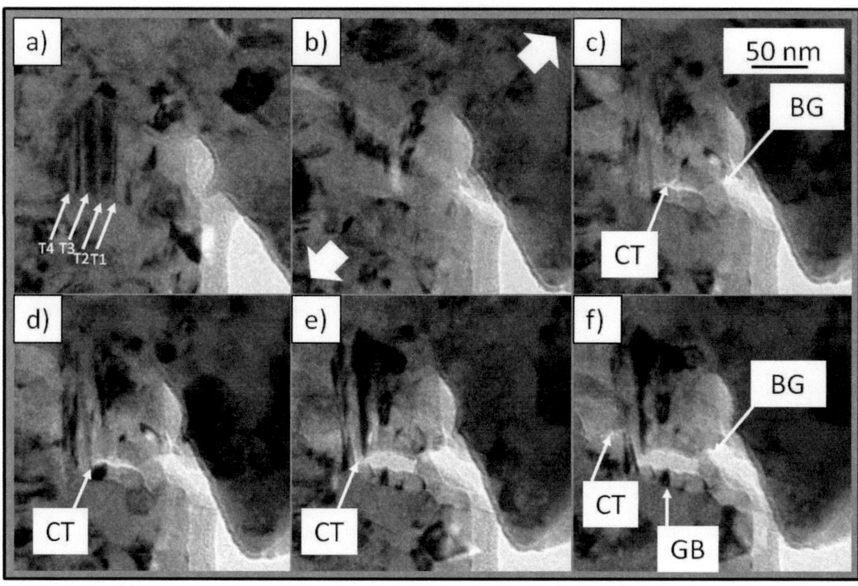

Figure 4-15: BF-TEM snapshots of single stages of fracture of sample ncCu62nmC5. Also the single twins in the grain next to the notch ground are labeled with T1-T4. Fracture starts in (b), progresses into the twinned grain in (c), where T1 (maybe also T2) is fractured. Then the crack propagated until (f). The applied loading direction is indicated with the white arrows in (b) as well as the crack tip with "CT", the GB of the nt grain with "GB" and the bridging grain at the notch tip with "BG".

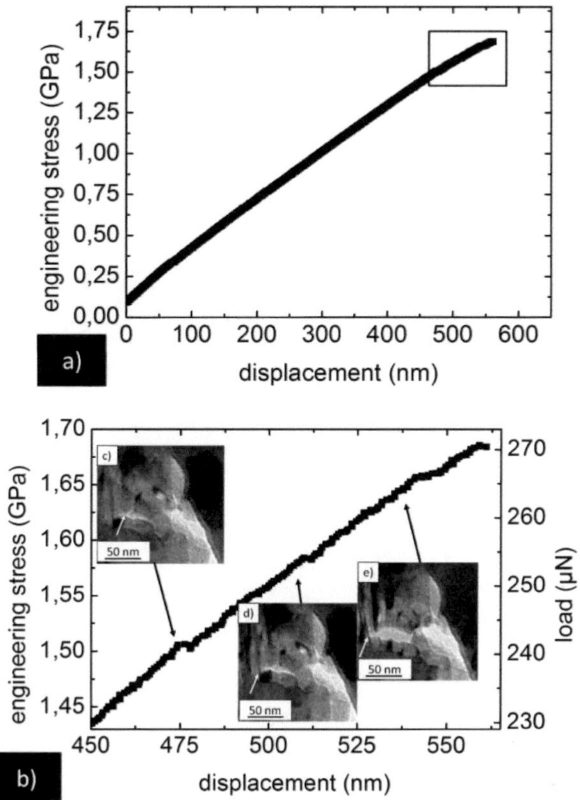

Figure 4-16: Engineering stress vs. displacement from the picoindenter of sample ncCu62nmC5 over the whole range (a) and the last hundred nm of deformation (b). The depicted region of (b) is marked in (a) with the black box. The letters (c)-(e) refer to the fracture events of single twins, which are depicted in Figure 4-15.

In the stress vs. displacement diagram, surprisingly high strains are reached. The deformation is basically linear, with slight plasticity at the end. At each twin fracture, some load drops appeared in the diagram, the positions are marked with small letters c-e, which correspond to the micrographs of Figure 4-15. The first load/stress drop at position c, can be referred to the fracture of T1 and T2. The load drops d and e refer to twins T2 and T3. The stress drops roughly 10 MPa. The stiffness of the PTP was already subtracted in this value, but the deformation of the PTP springs is included in the displacement applied by the picoindenter. Fracture started at the notch and the crack grew at 45° to the loading axis (see Figure 4-17). This is in contrast to the fracture morphologies of the other tests done in this project, where the crack grew normal to the loading axis (for example in Figure 4-9).

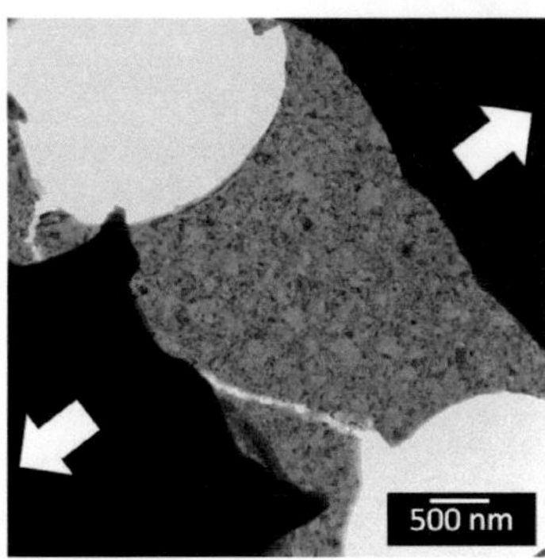

Figure 4-17: BF-TEM image of the fractured sample ncCu62nmC5 of Figure 4-15, where the diagonal fracture under ~ 45° to the loading axis.

The crack propagation in ncCu62nmC5 was determined from the micrographs in Figure 4-15 and was on the order of 250 nm. For the calculation of the created fracture surface dA, a thickness of 62 nm was used. The applied energy dW and elastically stored energy dE_{el} are determined from Figure 4-18.

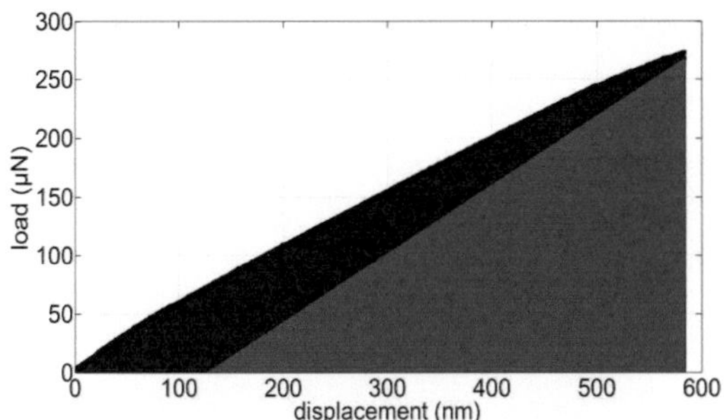

Figure 4-18: Load vs. displacement for the 47th cycle of ncCu62nmC5. The grey region represents the elastically stored energy, the difference between introduced and the elastically stored energy and the black region. This energy is available for crack growth.

The ratio of *dW and dA is* the energy release rate G_{IC}. For a stiffness of 56 GPa (as a mean between the two measured values of 41 and 71 GPa) a fracture toughness K_Q of 6.3 ± 1 MPa$\sqrt{m}$ can be calculated with Eq. (1). The used values are listed in Table 4-2.

Table 4-2: List of used parameters and determined fracture toughness in an energetically approach for an estimated crack length of 250 nm.

	a ~ 250 ± 22 nm
dW (10-11 Nm)	8.5
dE$_{el}$ (10-11 Nm)	6.2
dE (10-11 Nm)	2.3
dA (nm2)	31000 ± 3100
G$_{IC}$ (N/m)	746 ± 70
stiffness (GPa)	56 ± 15
K$_Q$ (MPa m$^{1/2}$)	6.3 ± 1.1

In post-mortem observation of the microstructure, intercrystalline fracture was found over the whole spectrum of grain sizes up to 140 nm. Transcrystalline fracture appeared only in grains bigger than 100 nm (Figure 4-19).

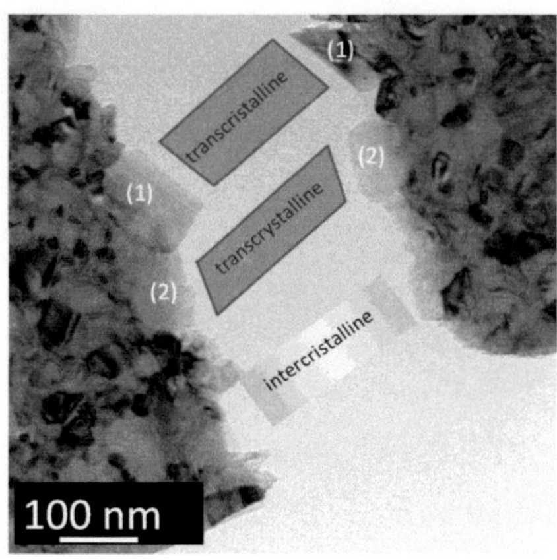

Figure 4-19: Intercrystalline and transcrystalline components of the fractured surface

4.5   Discussion

In this section, the presented results are critically discussed in terms of the experimental method. Then, the mechanical response of the monotonic and cyclic *in situ* deformation in the TEM is discussed and compared to literature values. For the monotonic tests, special attention is payed to the stiffness of small volumes of the nc Cu samples, their strength, as well as the high deformation to failure. During cyclic loading, the coarsening characteristics were a point of interest.

4.5.1 General aspects

One issue that comes along with small sample geometries is the maximal size of the defects, which might be in the range of the sample size [20]. But the probability that defects of critical size could be in the sample, becomes smaller. Those two points may lead to a higher deviation of the properties.

In terms of fatigue the effect of volume on the mechanical properties was already investigated [24]. There, the volume that is subjected to 90% of the maximal load was called risk volume. In conventional fatigue samples it is in the range of some mm^3 to around hundreds of mm^3 and a noticeable change in that range of sample dimensions was presented. In case of the *in situ* samples of this work with dimensions of 100 nm x 2 µm x 60 nm in the narrowest section the sample volume was ~ 10^{-11} mm^3. Accordingly the size and scaling effects on properties and mechanisms are expected to be significant, e.g. increased strength and change to GB dominated deformation.

Differences in the width of the grain size distribution, like it is obvious for sample ncCu62nmT1 and T3 (Figure 4-3 and Figure 4-7), might result from the deposition process. The samples were sputter deposited in several batches. Differences in the separate batches seem to be possible. The effect of the focused ion beam on the sample material can be neglected, as the beam is never focused on the tested region. Another possible consequence of FIB radiation is grain growth, depending on the angle of impact [111].

Besides GB, microvoids might be the only defects that are introduced during the deposition process and contribute to the defect density distribution. Very few stress concentrations like cracks should be created at the surface by the FIB-cutting process (Figure 4-9).

Another issue that comes along with the thin films is the exact determination of the films thickness in order to calculate the correct stress. Besides the desired thickness after the deposition process, the thickness was validated with SEM and AFM. Limiting factors in the SEM and AFM measurements is the resolution limit which has a significant contribution to the inherent error.

4.5.2 Mechanical aspects

The monotonic tests were conducted, to get an indication at what amplitude and to what stress levels the cycling would be meaningful. Another aspect was to measure the strength in such small volumes. The measured ultimate tensile strength (UTS) of ~ 2.5 GPa, is up to 3 times higher than measured by Gruber et al. [139], who also measured sputter deposited nc Cu films. It is 5 times higher than the value determined by Wang et al. [132] for their ball milled samples (severe plastic deformation) or inert gas condensated samples [140, 141]. The yield strength of nc materials was described to be 10-15 times higher as in its coarse grained counterparts [140]. All the samples had similar grain sizes. Nevertheless, ncCu62nmT3 had a wider grain size distribution with a fraction in the lower ufg regime (up to 150 nm). In comparison the grain size distribution of ncCu62nmT1 is purely in the nc regime (Figure 4-7). The microstructure that also contains bigger ufg grains accommodates more deformation and appears to be less brittle, like it was proposed by [36, 37] (see Figure 4-8). This explains higher strength of ncCu62nmT3 with 2.5 GPa instead of 1.6 GPa of ncCu62nmT1 to some extent. Furthermore, the small volume of the tested sample is no representative element of the failure distribution of that material leading to a large deviation of the measured values. Therefore, the results from these tests should not be generalized.

The yield strengths from the samples tested here are plotted in comparison to several values from literature in Figure 4-20. The production routes are not the same, but the microstructure, purity and grain size from the sputter deposited and inert gas condensated samples should be similar. In the monotonic loading of the as-deposited (as-dep.) samples, the tensile strength refers to the yield strength, as well as in the cycled samples in the last cycle.

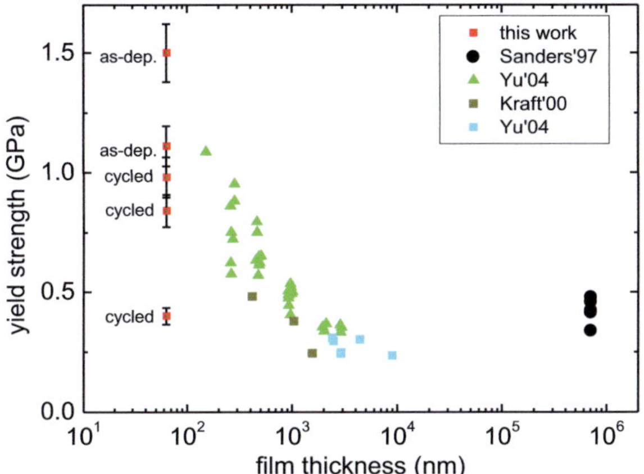

Figure 4-20: Yield strength of nc Cu thin films with a grain size of ~ 62 nm in comparison with literature values from Sanders'97 [140] (nc Cu), Yu'04 [142], Kraft '00 [143]. The samples from Yu et al. and Kraft et al had a grain size in the dimensions of the film thickness.

Figure 4-20 shows a strong increase of the yield strength with decreasing film thickness. Not just the film thickness of the samples tested in this work is of extreme small size, but also the width and length and therefore the sample volume of 10^{-11} mm^3. While the scaling and size effects come together in this size regime high stresses can be reached. At higher stress levels, also the scatter of those data points is increasing.

4.5.3   *In situ* testing

The *in situ* measurement showed strains of roughly 5-8 %, depending on the definition of the gage section. This is high for nc materials, which usually have strain to fracture of ~ 1 % [118]. An explanation for the unexpected high strain could be the small sample dimensions and as a consequence the strongly pronounced even strain state [20, 144]. At least for sample ncCu62nmT3, the wide grain size distribution, where the biggest grains are already in the ufg regime. Also the probability of fatal defects is much smaller in small sample volumes. In combination with the high strength, high stress levels can be reached, where dominant deformation mechanisms might change and usually brittle materials can develop some more plasticity.

In the yield strength vs. film thickness plot (Figure 4-20), the data points of the monotonic tensile tests on as-deposited material are well above the values of the cycled material. This indicates that the cycling has a weakening effect on the samples, since crack formation was not observed before the last cycle. Therefore the weakening of the material is related to microstructural changes during mechanical cycling.

The stiffness vs. film thickness and grain size is depicted in Figure 4-21 and Figure 4-22 in relation to literature values. The general trend is that stiffness is decreasing at smaller film thickness and grain size. The deposition methods of IGC, EB-PVD and sputter deposition are different, but the resulting microstructure should be similar in terms of purity and porosity. The samples described in [132] were produced by inert gas condensation (IGC). In the two tensile tests, a stiffness of 94 GPa (ncCu62nmT1) and 55 GPa (ncCu62nmT3) was calculated from the elastic slope in the stress strain curve (Figure 4-8).

Gruber measured a modulus of roughly 125 GPa (plane-strain) for [111] sputtered Cu thin films with a thickness between 80 and 800 nm in the bulge test [145]. The films were aged in order to exclude a grain size effect besides the film thickness effect. The excess volume and/or porosity are reduced during this heat treatment what explains higher values to some extent. The thinnest films were slightly thicker than the films used in this work, but the stiffness is larger by a factor of 2. This difference is attributed to the coarser grain size.

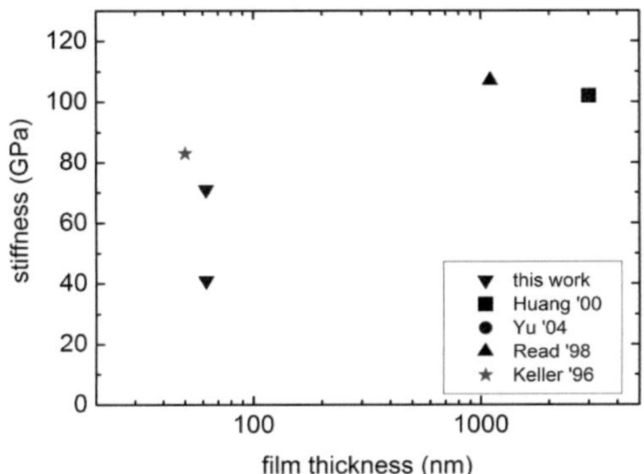

Figure 4-21: Stiffness vs. film thickness of the samples tested in this work and EB-PVD samples from literature. Values from Yu'04 [142], Huang'00 [146] (grain size ~ layer thickness), Read'98[147] and Keller '96 [148].

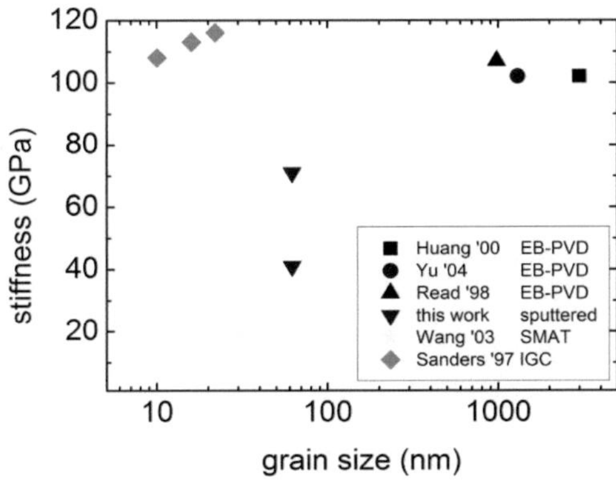

Figure 4-22: Stiffness vs. grain size of the samples tested in this work and values of samples from literature. Wang'03 (surface mechanical attrition) [132], Yu'04 (EB-PVD) [142], Huang'00 (EB-PVD) [146] (grain size ~ layer thickness), Read'98 (EB-PVD) [147], Keller '96 (EB-PVD) [148] and as bulk reference Sanders '97 (inert gas condensation) [140] (nc Cu)

The modulus or stiffness values on macroscopic samples from Sanders et al [140] were measured with the ultrasonic pulse-echo technique and are almost spot on the theoretical level for a isotropic modulus of 130 GPa. These values do not follow the trend that is indicated by the values determined from the thin film tensile tests. The sample deposition, heat treatment, preparation and dimensions as well as the testing technique have a distinct influence on the stiffness values.

Nc Cu was already discussed to have a modulus of values of a third of the coarse grained bulk modulus [91, 149]. There are several factors, which can lead to a reduced measured stiffness, for example:

- Testing geometry
- Texture
- Porosity
- GB processes like grain rotation and GB sliding

In the following section, these different factors that affect the thickness are discussed.

Depending on the assumed gage length (referential length for strain calculation), which might be shorter by a factor 0.75 (Figure 4-3) a corresponding lower stiffness can be expected, hence values of 71 GPa +/-10 % and 41 GPa +/-10 %.

In all utilized publications about electron beam physical vapor deposited (EB-PVD) Cu from [146], [150] and [142], the material had predominantly <111> texture but also a fraction of <100>. The nc Cu used in this work has a texture of mainly <111> oriented grains. A component of <100> oriented grains was stronger pronounced than in the bulk material (Figure 4-5). This is a factor that reduces the stiffness as the [100] orientation is weaker than the [111] orientation.

Low values of the stiffness might also be due to porosity in form of microvoids, which formed during deposition at the GBs what is visible in Figure 4-6 b. This is rather normal for sputter deposited materials. Furthermore, open porosity at the surface could have a big effect. At a sample thickness of 62 nm, the pores could locally reduce the thickness by ~ 10 %. Nevertheless, a roughness of 1.73 nm (RMS) was measured in the AFM. This value is roughly on the order of 2 % of the sample thickness and not expected to effect the mechanical response.

Another reason for a reduced stiffness might be the increased GB content. The grain boundaries might have a lower stiffness which would also reduce the overall stiffness of the whole sample [151]. Deformation processes like GB sliding and grain rotation might be activated at a lower stress and further reduce the stiffness. As a consequence, a small fraction of the external applied strain reached the grain interior and elastically changed the lattice parameter [152]. The fraction of GB related reduction of the stiffness must be on the order of 30% to explain the low stiffness of this work.

Grain growth of up to 30 % during tensile cyclical loading of nc Cu was observed before [91]. The fact that during cycling mainly the big grains coarsened and no coalescence in agglomerates was observed might be due to dislocation motion in these bigger grains. Homogeneous coarsening affected by GB related processes might then be more relevant during cycling.

In sample ncCu62nmC5, no pure mode I loading seems to be applied, as the right flank of the notch in the images of Figure 4-15 is moving towards the tip of the notch what is an indication for a superposition of a tensile and a bending mode. Reasons for mode mixity could be a damaged PTP device, with different stiffness of the springs on both sides. A contact point of the indenter tip, too far off the middle would have a similar effect and induce a bending moment on the sample. The actual reason is also expected to be a result for the fracture under 45°. The high loading rate (140 nm/s compared to ~10 nm/s) is not expected to have an influence on the global fracture morphology, as the fracture grew more under 45° instead of normal to the loading axis like in the low rate experiments. The stress/load drops in Figure 4-16, due to fracture of single twins are on the order of 0.7 µN. The first load drop is referred to the whole cross section of 62 nm x 2587 nm what corresponds to a stress drop of 4.3 MPa. If the load drop of 1.5 µN is referred to the cross sectional area of a twin (62 nm x 15 nm = 930 nm^2), a stress of ~ 0.8 GPa can be calculated.

The grain, right at the bottom of the notch is not nanotwinned. It is weaker and therefore more receptive for deformation than the nt regions next to it. This can be observed between c and f. The crack is already beyond that grain and in the nt region, while this grain is still of integrity (this grain is labeled with "BG" in Figure 4-15). It follows the opening displacement of the flanks of the crack, but connects them until final rupture of the sample. Until that point the strain that was accommodated by that grain is on the order of 20 %. Before fracture of the last twin of that nt grain, the crack seems to be already propagating into the next grain (see Figure 4-15 b). This is commonly known as bridging and e.g. was presented for bio materials by Nalla et al. [153] and for ceramic composites by Kaute et al. [154]. The underlying mechanism is described as the fibers or ligaments are pulled out of the surrounding matrix during the crack opening process which provides enhanced fracture toughness. Friction between the strong fibers and the matrix is dissipating energy and causing the increase in crack resistance [155]. A different bridging mechanism has to be prevalent in this work as the material is not a composite and no ligaments are pulled out of the matrix. Instead, twin ligaments or "soft", e.g. preferential crystal orientation, untwinned grains are the bridging elements and are strained between the crack flanks. This leads to an increase of the crack tip radius and might therefore influence the crack propagation to some extent. Due to the small size of the involved volume the total dissipated energy and finally the impact on crack resistance are not expected to be relevant. In larger samples, where more plastic deformation could be dissipated, a more effective crack bridging and toughening might be possible. The fracture toughness for nt microsamples are reported in literature to be at least on the order of 15-20 $MPa\sqrt{m}$ [54] and therefore by a factor of 3 higher than the values of the here presented nanosamples. Another mechanism during crack growth in the nt material could be crack blunting [156]. During crack growth dislocations are nucleated within the vicinity of the crack tip and interact with TBs. This can lead to a trapping of those dislocations and increase crack growth resistance.

In contrast to ncCu62nmC6, cracking of ncCu62nmC5 looked much more elastic without plastic deformation, hence a value for fracture toughness is tried to be determined in an energetic approach. The fracture toughness K_Q = 6.3 $\pm$ 1 MPa$\sqrt{m}$ seems to be too high to be in acceptable agreement with a value of 3.8 MPa$\sqrt{m}$ for nc Cu from Keller et al. [148]. Reasons might be an overestimated critical energy release rate, due to an underestimated crack length as a result of crack path tortuosity which is not necessarily visible in the TEM image. But also the plane strain conditions in the thin sample might account for high values of fracture toughness. Probably the biggest issue is the size of the plastic zone r_p which is defined by Eq (3-2) after [74]. With a fracture toughness of ~ 5 - 10, and a yield strength of 1.5 GPa, it can be estimated to be on the order of r_p ~ 1 μm. Following the ASTM norm [157], the sample should be twenty times bigger than the size of the plastic zone in order to be out of the regime of large scale plasticity. This is not the case and therefore this is another argument that the used value cannot be used as a valid fracture toughness K_{IC}.

A poisson ratio of 0.3 was used as the determined 0.03 did not seem to represent the real cross contraction, as it was hindered by the surrounding material. Real cross contraction is not expected to occur perpendicular to the width of the sample, where it could not be measured.

Figure 4-14 visualizes that the length of the upper edge of that grain and the grain diagonal increased during loading. Some plastic deformation is noticed in the upper part of the grain, as the length stays constant after the loading. After the 9th cycle, the crack didn't really change in terms of length, but slightly in terms of width.

In ncCu62nmC5 fracture occurred in the nt grain roughly 10-20 nm next to the GB. While in the slightly bigger central grain of ncCu62nmC5 the crack grew along the GB.

4.5.4 Mechanisms

Interesting for the analysis could be processes like GB decohesion (or grain rotation), which would refer to the competing mechanisms. That would result in more or less brittle or ductile behavior, depending on which mechanism is dominant. Also of interest are strain procedures in terms of local strains or global strains, which can both give an understanding of ductility and lateral contraction. Also crack mechanisms like crack formation and crack growth, could be quantified by the crack tip opening.

Intercrystalline cracking is expected to be easier at low grain sizes. In bigger grains transcrystalline fracture needs less energy and is more likely.

4.6 Summary

Our TEM observations on fatigued nc Cu samples have displayed the existence of single dislocation entanglements instead of dislocation structures like in bulk materials. This is similar to what has been found in fatigued thin film samples [110] with a film thickness or grain size smaller than 400 nm.

4.6.1 Monotonic tensile loading

1) The tensile test showed, that nc materials can be tested *in situ* reaching remarkably high flow stresses of up to 1.5 GPa and fractures after straining to 5 % and locally of up to 8 % in the gage section, reaching a UTS of up to ~ 2.5 GPa. Comparison with literature shows that these values are higher than expected from larger thin film samples which was explained by plastic deformation constraints due to the smaller sample volume. Therefore, the small volumes provide a reduced microstructural capacity to support active deformation mechanisms, e.g. dislocation mediated plasticity.

2) A Stiffness of 55 GPa and 94 GPa were measured. Reasons that can cause the low stiffness values are the condition of the sample like porosity and the testing technique, and the grain size.

3) Analysis of the strain with DIC is possible in regions of no/limited deformation, but hard to accomplish in the middle of the sample due to contrast changes.

4.6.2 Effects of cyclic loading

4) Cracking in a notched sample at low strain rate (5 nm/s) was observed along a GB until it stopped there. A big amount of plastic deformation

seemed to be accommodated, right at that neighboring grain by shearing of the upper part. At high strain rate (140 nm/s) the fracture looked quite different. The crack grew at 45° to the loading axis and even a nt grain in the region of the notch tip seemed to fracture intercrystalline.

5) In the cycling tests, grain coarsening formed to a more bimodal grain structure.

6) Grains with a size of up to 140 nm tend to fail intercrystalline, while grains larger than 100 nm show more transcrystalline failure and seem to undergo shear thinning.

7) A fracture toughness K_Q of roughly $6.3 \pm 1\ MPa\sqrt{m}$ was determined by an energetic approach, what is higher as the value measured by [145].

4.7   Acknowledgement

Thanks are given to John Turner, Dr. Hua Guo and Doreen Ah Tye from NCEM for their help with the TEM. Dipl.-Ing. Paul Vincze for the AFM measurement, Dipl.-Ing. Ralf Witte and Dipl.-Ing. Jochen Lohmiller for the XRD measurements as well as Dr. Christian Kuebel and Dr. Patric Gruber for productive discussions. The authors acknowledge support of the DFG (SFB499N01) and the National Center for Electron Microscopy, Lawrence Berkeley Lab, which is supported by the U.S. Department of Energy under Contract # DE-AC02-05CH11231.

5 Deformation and microstructural stability of nt Cu

5.1 Abstract

Monotonic and cyclic tensile tests were conducted on magnetron sputtered nanotwinned (nt) Cu thin films (thickness of ~ 20 µm), in a small scale tensile testing setup. nt metals are a promising class of nanostructured (ns) materials with high hardness and strength following the Hall-Petch-relation. Furthermore, nt metals do show moderate ductility. Face-centered cubic metals with low stacking fault energy (SFE) like Cu have the tendency to develop this kind of nt structure. As the twin boundaries (TBs) are energetically lower than grain boundaries (GBs) they are thermodynamically more stable as their nanocrystalline counterpart while under mechanical load they still can become unstable. We used TEM to quantify the microstructural changes of detwinning that appeared during monotonic and fatigue testing. It appears that the detwinning process during tensile and fatigue loading leads to local softening. The shear deformation is therefore accommodated in a very small band of roughly 15 µm across the 500 µm wide sample.

5.2 Theory

5.2.1 Properties and microstructure

Nanotwinned (nt) materials with a twin spacing λ of less than 100 nm show similar Hall-Petch strengthening characteristics [49, 158, 43, 159] and strain rate sensitivity levels [158, 54, 63], as their nanocrystalline (nc) counterparts, while they are more ductile [43]. The reason for high ductility of 10-20 % [159,

118, 160] or even 30 % [43] are the twin boundaries (TBs) which act as dislocations sources [158, 63]. Dislocations can be emitted from stress concentration in TBs, which was shown by *in situ* TEM [56]. The dislocation density can be up to two orders of magnitudes higher than in coarse grained (cg) Cu [63]. At the same time, these TBs act as barriers to dislocation motion [43, 63].

With a low stacking fault energy (SFE) [63], the epitaxial grown films of Cu form nanotwins separated by Σ3 (111) coherent twin boundaries (CTB) and columnar domains separated by Σ3 (112) incoherent twin boundaries (ITB) [161]. Owing to the surface of the etched Si (110) the epitaxial Cu films grow in a preferred orientation with a negligible in-plane misorientation between domains build of stacks of nanotwins. As a consequence, the film is single crystalline [42].

5.2.2 Deformation behavior

Sputtered nt Cu has a low density of active dislocations [162], which are free to move, as they are embedded in ITBs. To mobilize these dislocations or nucleate them from the domain boundary (DB), the load has to be increased. Once they are nucleated and mobile, the necessary stress to sustain the deformation can decrease, which leads to a yield strength peak [137]. The plastic strain rate is given by the Orowan equation

$$\dot{\varepsilon} = |\vec{b}|\rho v \qquad (5\text{-}1)$$

where $|\vec{b}|$ is the Burgers vector, ρ the dislocation density and v is the average dislocation speed. A low dislocation speed v leads to higher necessary stress levels and thus sharper and denser populated glide bands. Depending on initial dislocation density a yield peak might form during loading, if there are not enough dislocations to keep up with the deformation. In contrast, the yield peak does not occur if the dislocation density is increased by a process like cold rolling [162].

5.2.3 Microstructural evolution

Another aspect of nt materials [163] is the enhanced fatigue life and endurance limit compared to cg Cu. One major aspect is that the microstructure influences the path of crack growth. Nc materials tend to have almost no mechanisms for crack deflection which lead to the case that less energy is necessary to force the crack to propagate [88]. That is different in the nt material with often ultra fine columnar superstructure of stacks of nanotwins. The crack deflection might appear at the ITB, as the domains are much bigger than the grains in nc material. This leads to a high crack tortuosity and forces the crack to follow energy consuming paths [54]. Former fatigue tests on nt Cu pointed out that especially the fatigue strength in the LCF regime is higher than in their coarse grained counterparts. Presumably, this is due to increased strength and at the same time preserved ductility [163]. During mechanical loading, the nt structure which providing beneficial properties, can be dissolved. This process of coarsening is referred to as detwinning [164, 165]. It might be one of the processes in nt Cu leading to softening and localization of deformation and finally failure [57].

In the next sections, the tensile and fatigue testing procedures on the nt Cu samples are presented, followed by the results. The microstructure and its changes were observed with TEM techniques. In the last part, these findings are critically discussed.

5.3 Experimental

Epitaxial Cu (111) films were magnetron sputter deposited with a thickness of 20 µm on HF etched Si (110) substrate [160] at the University of Texas by the group of X. Zhang.

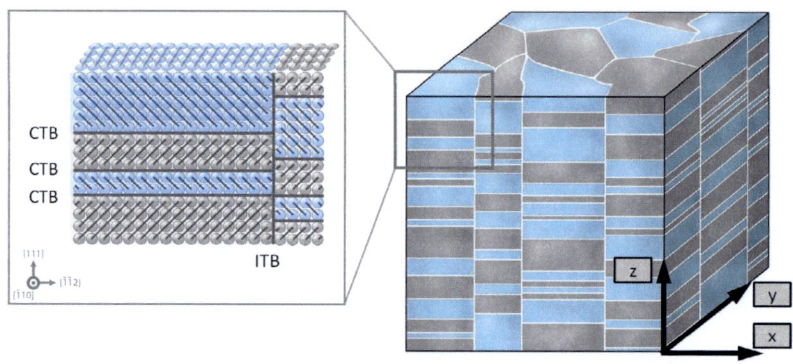

Figure 5-1: Schematic drawing of the domain and twin structure of the cross section of a sputtered nt Cu film. The orientation of the sample is as follows, x in loading direction, y is the second in-plane orientation and z is the out of plane orientation.

The films were cut with a diamond wire saw into small rectangles, which had slightly larger dimensions than the samples. As a sample geometry dog bone shapes were chosen with a length of 4.72 mm and a width of 0.50 mm in the gage section. The shape was inspired by the work of Hemker et al. [166], and optimized by the tension triangle method [167]. The optimized shape (Figure 5-2) prevents stress concentrations at the transition region between the grips and the gage section. The nt Cu platelets were stacked and cut into the desired dog bone shaped geometries by the use of a wire electron discharge machine (EDM AGIE 100 –cut, Techma, Germany). The EDM current used for the machining was around 8 A. Additionally samples were cut in a LASER with a power of 18 W and a puls length and frequency of t = 100 ns and f = 60 kHz respective-ly resulting in a cutting gap of roughly 30-40 µm.

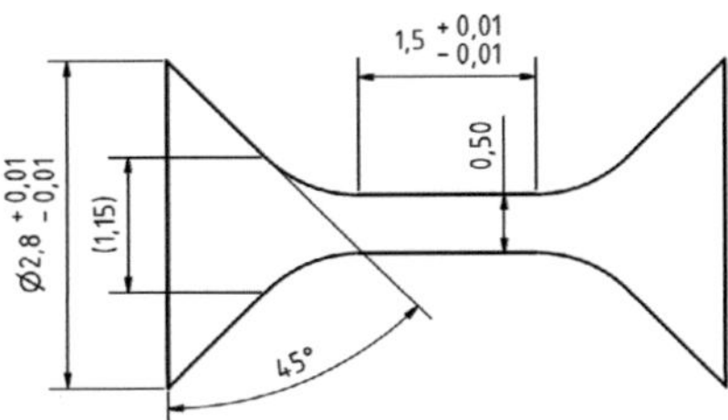

Figure 5-2: Technical drawing of the dog bone shaped sample geometry used for the tests on nt Cu microsamples. The shape of the transition between gage and ear section was optimized by the tension triangle method to prevent stress concentrations [167].

The displacement controlled tensile and load controlled fatigue tests were conducted in a custom built setup (Figure 5-3). It consisted of a piezo actuator with an integrated sensor for positioning (60 µm driving range, resolution of 1.2 nm, Physikinstrumente, Germany).

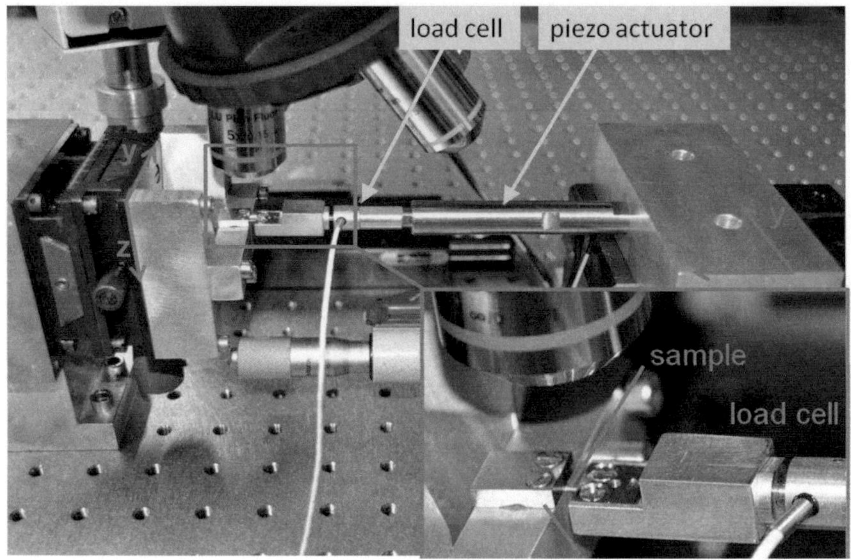

Figure 5-3: Custom built tensile testing setup, consisting of a piezo actuator, the load cell, to grip the samples and the microscope. (private communication T. Kennerknecht)

For load measurement, a dynamic load cell ($\pm$ 50 N, resolution of 25 mN, Disynet, Deutschland) was used. It was fixed to the piezo on one side and carried the grips on the other. The piezo actuator as well as the sample holder was mounted to x-y-stages in order to align the sample properly. Image of the sample were captured during the tests through the microscope. This allows non-contact strain measurement by digital image correlation (DIC) in a post process. The matlab® code uses fiducial markers that remain at the surface with a high contrast difference and follow their displacement in order to quantify the deformation [138].

The following labeling convention for the samples is used: The name of each sample consists of the material (nt Cu), the film thickness in nm, the testing mode (T for tensile, LF for low frequency and HF for high frequency fatigue testing) and a ongoing number.

Scanning electron microscopy (SEM) analysis was conducted in a dual focused ion beam (FIB) (FEI Nova 200, Hillsboro, Oregon, USA). The transmission electron microscopy (TEM) lamellas were prepared at 30 kV and 20 nA, cleaned at 100 pA at +/- 1.5° tilt angle and then polished at 5 kV with a beam current of 70 and 10 pA at +/- 5° tilt angle. For the analysis a FEG TEM (Tecnai F20 , Philips, Netherlands) was used at an acceleration voltage of 200 kV. In TEM mode the camera length was set to 8 cm, while a spot size 8, a condenser aperture of 50 µm and a 4.5 kV FEG was used. The TEM was equipped with a µSTEM for orientation mapping (OIM, Nanomegas, France) that was run with 100 fps.

The crystal orientation at the surface was measured by the use of an EBSD (Oxford Instruments, United Kingdom) at 20 kV and 6.6 nA with a step size of 50 nm. To remove the oxide on the sample surface, the specimens were etched with a solution of 90 % ethanol and 10 % nitric acid for a few seconds. The micro structure was quantified by the use of a custom written matlab® code in a line intersection method [106].

5.4 Results

5.4.1 Tensile test

The lines that are visible on the surface of ntCu20µmT6 in Figure 5-5 and on ntCu20µmLF7 in Figure 5-7 are not related to the shear deformation. They already existed before the test for example on the platelets right before the EDM cutting process (see Figure 5-4).

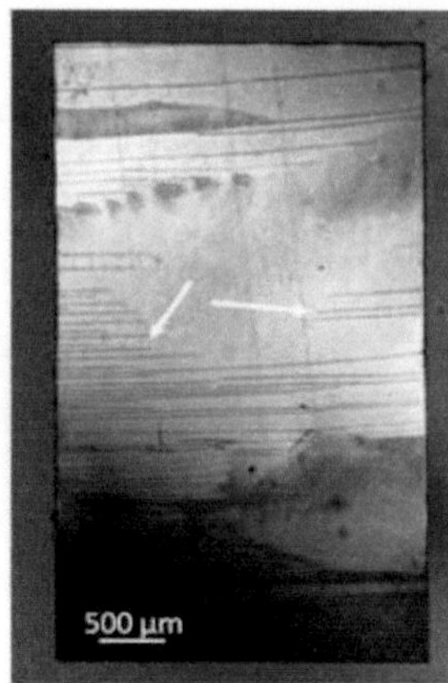

Figure 5-4: Light microscopic image of platelets of nt Cu. Parallel lines on the surface that are related to the deposition process are marked with white arrows.

The main deformation mode of the tensile tested samples was shearing of under ~ 45 °, as is clearly visible in the samples ntCu20µmT5 and ntCu20µmT6 in Figure 5-5.

Figure 5-5: Light microscope images of the tensile test sample ntCu20µmT4 and ntCu20µmT5. Sample ntCu20µmT5 sheared in two directions, to compensate the offset of the samples during loading. The loading direction is vertical. (private communication T. Kennerknecht)

Typical stress vs. strain curves for ntCu20µmT4 and ntCu20µmT5 can be seen in Figure 5-6. After yielding at ~ 535 MPa and ~ 555 MPa the samples start to deform homogeneously by shear and are gliding off at about 2.5 % strain. No strain hardening and not much yield is recorded before reaching the ultimate tensile strength (UTS). Afterwards, the strength reduces due to the decrease of the cross section. The samples start to glide off on a 2nd shear plane at about 4 % total strain. At that point, the cross section is further reduced as well as the engineering flow stress. Surprisingly, ntCu20µmT5 undergoes a large elongation of up to 10 % and shows a totally different slope when the 2nd shear plane is activated. The analysis of the slope in the elastic region of the stress vs. strain curve of Figure 5-6 provides a Young's modulus of 92 GPa. Typically, the Young's modulus is not measured in that regime but in unloading segments after the first plastic deformation occurred. From such an unloading segment in ntCu20µmT4 a value of up to 140 GPa was measured.

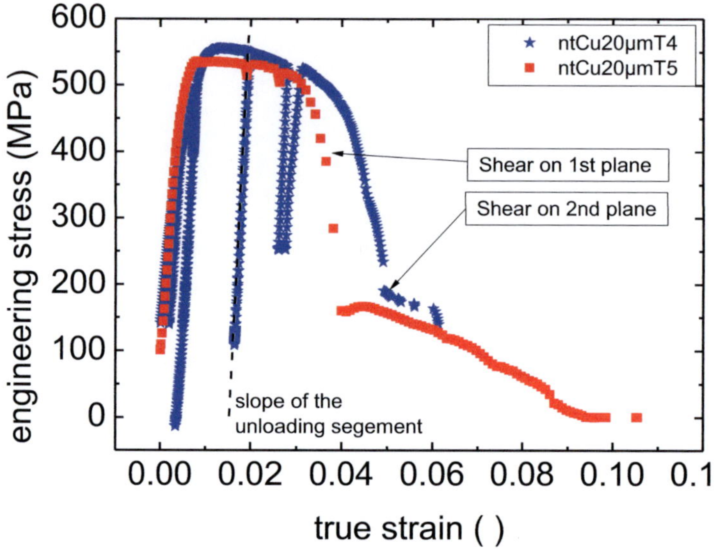

Figure 5-6: Stress strain curve of tensile loading of ntCu20µmT4 and ntCu20µmT5 (private communication T. Kennerknecht)

Several specimen and testing parameters of the load controlled tensile tests are summarized in

Table 5-1.

Table 5-1: Sample geometry and results from the tensile tests on nt Cu (private communication T. Kennerknecht).

sample	h (µm)	width (µm)	$\dot{\varepsilon}$ (1/s)	σ_y (MPa)	UTS (MPa)
ntCu20µmT4	20	500	$\sim 2*10^{-4}$	~ 400	555
ntCu20µmT5	20	500	-	~ 500	535

5.4.2 Fatigue test

The fatigue tests were done in two different frequency regimes, at low frequency (LF) $f \approx 0.5$ Hz and a high frequency (HF) $f = 50$ Hz respectively. The tests at a frequency of less than 1 Hz had a hold segment at the maximum load. For the samples tested at high frequencies shear bands are not clearly visible in an optical microscope due to the shallow width. The sample that did not fracture ran into an abort value of a load drop of 3.9 %. Not all samples were fatigued until fracture occurred. NtCu20µmLF7 and ntCu20µmLF8 were fatigued at a load amplitude of ~ 50 % of the UTS and showed necking (see Figure 5-7).

The sample lifetimes are shown in a Woehler-curve (SN-plot) (Figure 5-8). The nt Cu samples clearly experience longer fatigue lives when tested at a frequency of 50 Hz compared to frequencies below 1 Hz. Due to the low number of conducted tests in a range between $5*10^2$ and $2*10^5$ cycles and a load amplitude of 190 and 270 MPa, the slope of the corresponding regime cannot be determined. Therefore it is not sure if the samples reached the HCF regime and if the fatigue lifetime was defined by ductility or strength, respectively.

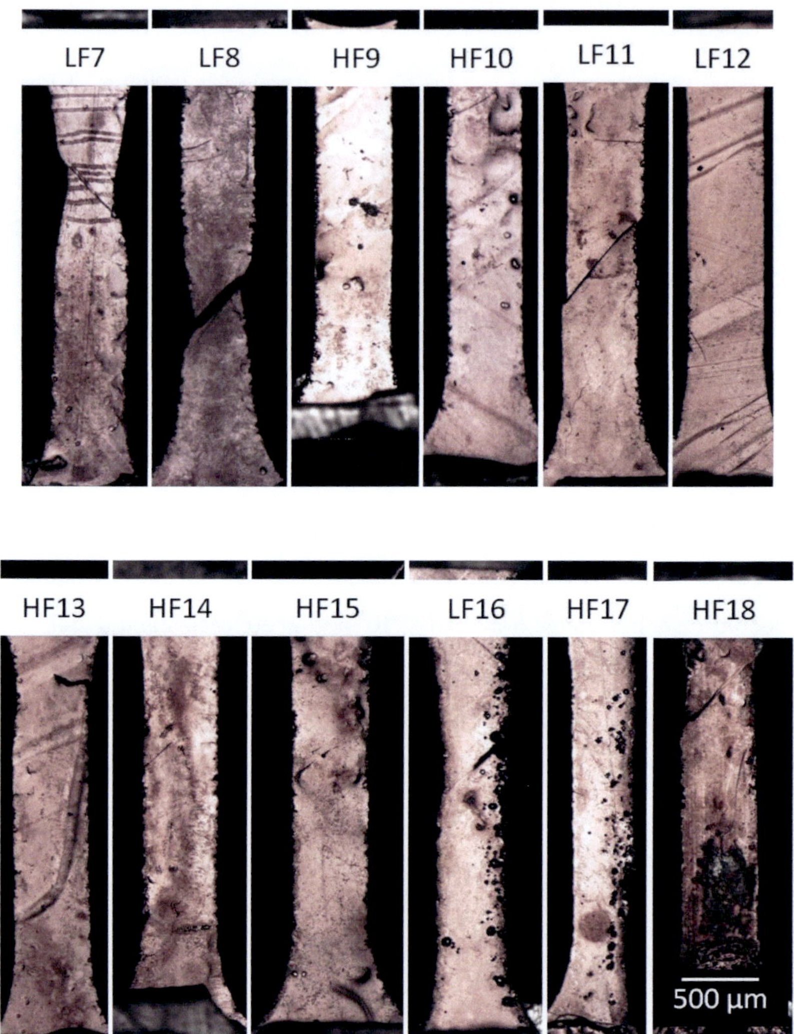

Figure 5-7: Light microscopic images of the fatigued samples ntCu20µm LF7, LF8, HF9, HF10, LF11, LF12, HF13, HF14, LF15, LF16, HF17, HF18. The loading direction is vertical.(private communication T. Kennerknecht)

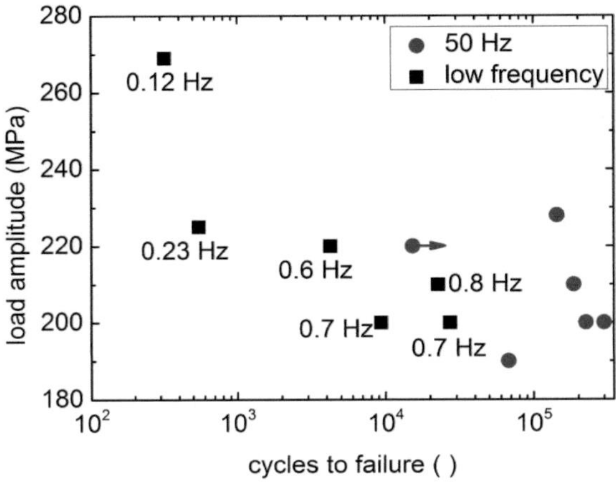

Figure 5-8: SN-plot of the tested nt Cu specimen. Clearly visible is the longer fatigue live of the tests conducted at higher frequency (50 Hz) (private communication T. Kennerknecht).

The testing parameters and results of the fatigue tests conducted on nt Cu are summarized in Table 5-2. At 50 Hz only the load amplitude was varied. At low frequencies (< 1 Hz) the load ratio, the frequency (loading rate) and the length of the hold segment were varied. N_f

Table 5-2: Cycle parameters and resulting fatigue life times. The specimen tested at 50 Hz are labeled with an "HF" at the end of the name. Samples tested at a frequency < 1 Hz with "LF" (private communication T. Kennerknecht).

Sample	ΔF (MPa)	N_f	R	f (Hz)	Hold segment trapez (s)	Loading rate (N/s)	v (mm/s)
at 50 Hz							
HF9	228	1.43E+05	0.1	50		400	
HF10	220	1.53E+04	0.1	50		400	
HF13	210	1.86E+05	0.1	50		400	
nHF14	200	3.00E+05	0.1	50		400	
HF17	200	2.25E+05	0.1	50		400	
HF18	190	6.75E+04	0.1	50		400	
at low frequency							
LF7	225	547	0.02	0.23	2	26	??
LF8	269	320	0.16	0.12	1.97	2.4	0.007
LF11	220	4.25E+03	0.1	0,6	0.1	7,8	0.02
LF12	210	2.26E+04	0.1	0,8	0.1	7,8	0.02
LF15	200	3.34E+03	0.1	0,7	0.1	6,5	0.02
LF16	200	2.72E+04	0.1	0,7	0.1	5,2	0.02

5.4.3 Fractography

The final deformation appeared to happen in a rather straight and shallow shear band at roughly 50° to the loading axis. The affected volume of the localized deformation was limited to a narrow band (Figure 5-9). The shear distance of 55 µm on a sample diagonal of 666 µm along the shear line relates to 8% (Figure 5-9 a). At an estimated shear band thickness of 15 µm the ratio to the shear diagonal is on the order of $\gamma = 370$ %.

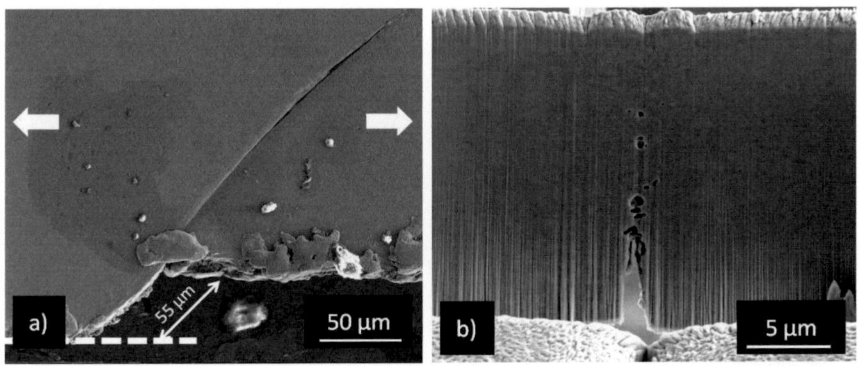

Figure 5-9: SEM micrograph of ntCu20µmLF11: (a) the sheared offset of about 55 µm was accommodated in a roughly 15 µm wide band. The loading direction is indicated by the white arrows. (b) The heavily deformed shear band has a thickness of around 10 µm. SEM image of the cross section of ntCu20µmLF11 with coarsened regions of a few microns around the shear band, which is located in the middle of the image along the wholes.

Besides a little neck in plane direction, the surface of the sample does not seem to be affected by fatigue. No indications for dislocation activity and the formation of extrusions could be discovered. The size of the heavily deformed zone, where detwinning occurred can be estimated from (Figure 5-9 b) to be on the order of 15 µm. In this FIB cross section with the shear band in the center, the detwinned region appears to be much coarser than the nanotwinned regions away from the shear band.

In all fracture observations, shear deformation played a dominant role. In the fully separated fracture surface of ntCu20µmLF8, dimples of a size of several microns were observed. They are one order of magnitude bigger than the initial twin domain size (Figure 5-10 a). The fracture surface of ntCu20µmLF11 which formed a shear band but was not fully separated, has flat features with characteristic sizes in their thickness and width. The width of these features is roughly 1-2 µm, while the thickness is a few hundreds of nanometers. The features have the same orientation as the twins but are roughly ten times of the initial twin spacing of ~ 25 nm (Figure 5-10 c).

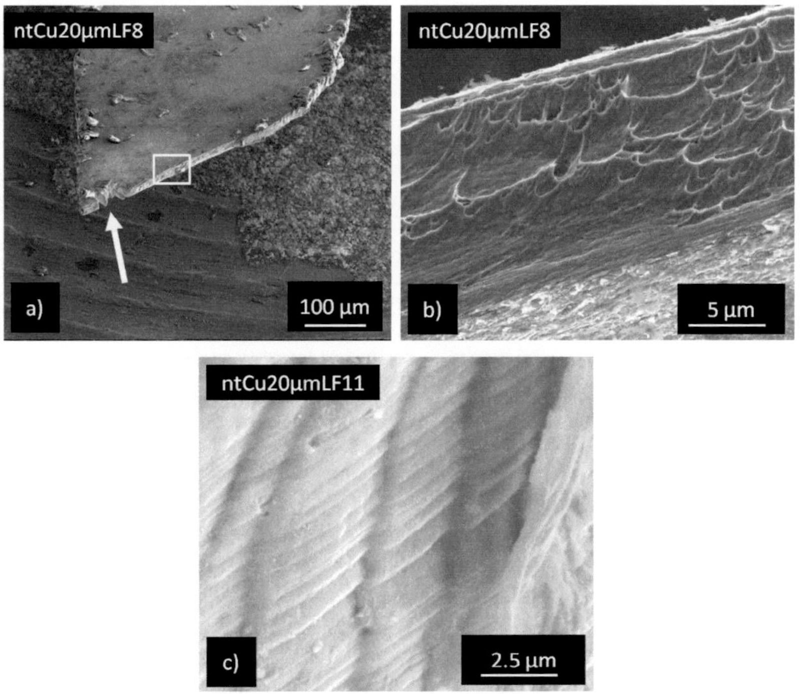

Figure 5-10: SEM micrographs of the fracture surface of the fatigued sample with (a) sample ntCu8LF and the location of the extracted TEM lamella is marked with the white arrow and (b) separated ntCu8LF with dimples in the size of one order of 10 magnitudes larger than the initial twin size of a few hundreds of nm at the location framed in (a). (c) fractured ntCu20µmLF11 and surface features with the same orientation as the initial twins.

5.4.4 Microstructure

To get an impression of the microstructural changes that arise due to the local heating from cutting, i.e. by EDM and LASER, cross sections of the edge of the sample were cut in the FIB. The heat affected zone is visible by a larger twin spacing directly at the surface. In the case of the EDM process this zone is on the order of 5-8 µm whereas it is only ~ 2 µm for the LASER cut samples (Figure 5-11).

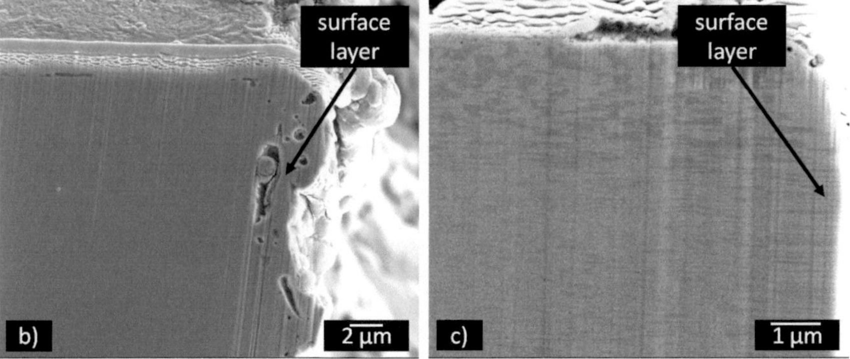

Figure 5-11: (a) Orientation map of the as-deposited twin structure, containing CTB and ITB. SEM micrographs of the heat affected zone after cutting in (b) EDM and (c) LASER.

An additional issue with the EDM process is the formation of porous and brittle oxide and hydroxide on the processed surface, which can act as stress concentrators [168]. The surface of the LASER cut samples shows less roughness. After this aspect of sample preparation, the general microstructure and its evolution are going to be presented. Orientation mapping [169, 170] in the TEM gives further insight to the overall microstructure of the as-deposited material. It consists of a (111) columnar ultra fine domain superstructure with a width of a few hundred nm and the nt structure within. Several twins seem to reach across DBs (Figure 5-11 a). Clearly visible is that there are only two colors in the orientation map, referring to the two kinds of differently oriented modifications (twins). The twin size distribution has a maximum at ~12 nm and twins have a spacing of up to 80 nm.

An EBSD-mapping of the in-plane surface additionally indicates the microstructure to consist of (111) planes perpendicular to the growth direction (Figure 5-12 a). An indexing rate of ~ 50 % was reached. Two types of orientations were found which refer to the two orientations of twin modifications (Figure 5-12). The pole figures of the (c) {111}, (d) {110}, (e) {100} and (f) {112} planes characterize the surface near domains.

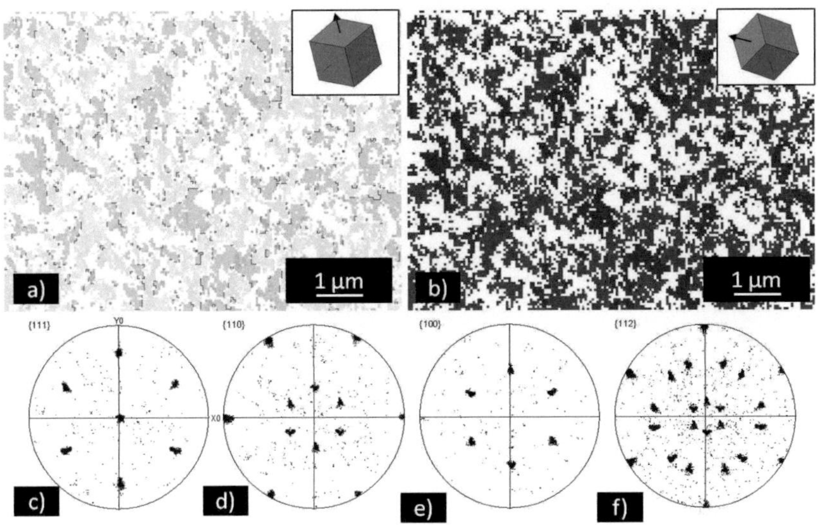

Figure 5-12: (a and b) EBSD mapping of an undeformed as deposited nt Cu sample, indicating the two existing orientations, twin modifications respectively. (c-f) Orthogonal projection maps of several orientations demonstrate the single crystalline character of the material.

The DBs that separate the single layers of twins are built up by aligned ITB along the growth direction (Figure 5-13 a and b). Typical for the sputter deposition of materials are pores in the size of tens of nanometers along these DBs.

In the deformed samples (Figure 5-13 c), the ITBs seem to be shifted in the one or the other direction. The former ITB position is marked by the pores aligned along the DBs. During loading, ITBs are separated from the domain walls and therefore from the pores.

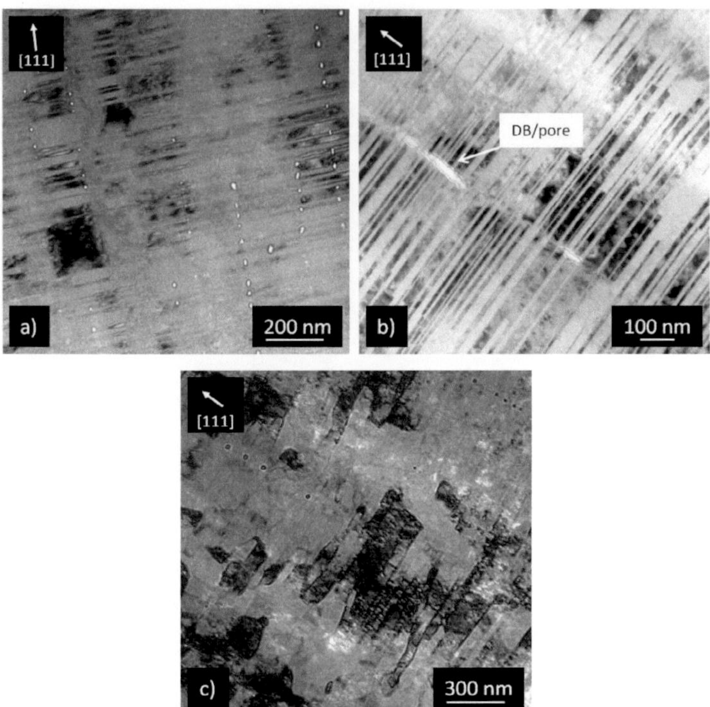

Figure 5-13: BF-TEM micrographs of (a) initial nt Cu with pores at the DBs and (b) clearly visible elongated pores in ntCu20µmLF11, which get clearer when viewed in slight underfocus. (c) Pores that are an inherent component of the sputtered material seem to be separated from the DBs, where they initially belonged to in the fatigued nt Cu.

The detwinned regions of the samples ntCu20µmT5, LF7, LF8 and LF11 are presented in the following section.

The sample ntCu20µmT5 was deformed up to the point where it fractured plastically under tension. To extract a TEM lamella, a Pt bar was deposited perpendicular to the sheared surface at the edge of the sample. This lamella is shown in Figure 5-14 a and b, where the little notches at the upper and lower part of the sample indicate the edge of the Cu sample next to the Pt. From that point on, the microstructure seems to be detwinned partially until a distance of at least 9 µm. In the very right electron transparent window, the nt microstructure seems to have its initial feature size (Figure 5-14).

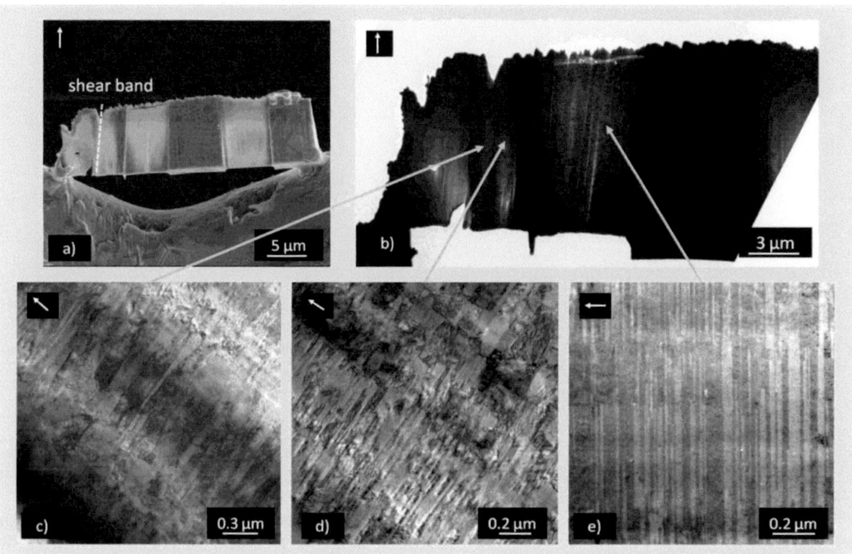

Figure 5-14: (a) SEM micrographs of ntCu20µmT5 with the ruptured surface on the left side (see arrows)and (b) TEM bright field images at three locations on the sample (c,d,e). The angle of rotation is different due to different magnifications in TEM. The (111) orientation is marked with a white arrow in each micrograph.

The cyclic loading of sample ntCu20µmLF7 was stopped before separation. Figure 5-15 a visualizes the rough EDM-surface, the strong necking on both sides at the sheared zone and where the TEM lamella was extracted.

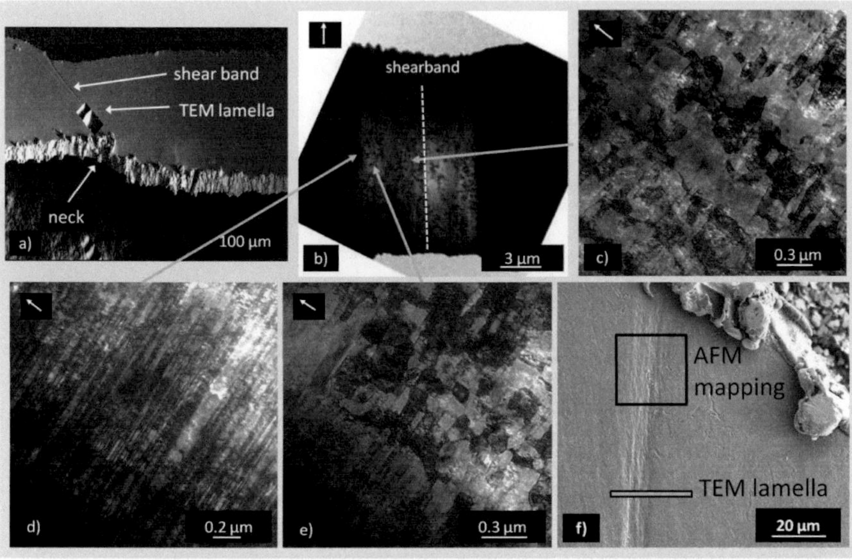

Figure 5-15: BF-TEM and SEM micrographs of the fatigued sample ntCu20µmLF7. (a) SEM micrograph showing the shear band and the rough surface from EDM processing. (b) BF-TEM micrograph of the sample with the shear band. (c, d, and e) are BF-TEM micrographs on different position with totally detwinned and ufg microstructure, nt and partially detwinned. f) SEM micrograph of a sheared region at the edge showing the position, where the AFM mapping was conducted and the lamella extracted across are marked with black rectangles. The (111) orientation in the BF-TEM micrographs is marked with a white arrow in.

The depth of the shear band was measured with four profile lines by an AFM mapping to be on the order of 400 to 600 nm (see Figure 5-16 a and b)

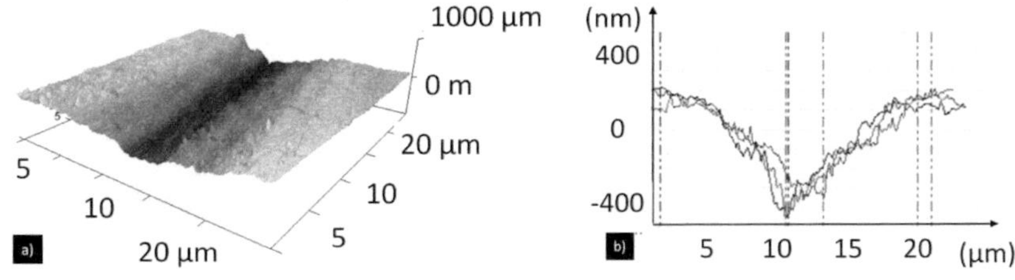

Figure 5-16: (a) AFM mapping of the middle of region of Figure 5-15 f. (b) 4 profiles at different positions of the AFM mapping

The TEM lamella was extracted in a way that the shear band with its detwinned microstructure is in the center of the lamella (Figure 5-15 a). Symmetrically on both sides follow the transition region with a partially detwinned microstructure (Figure 5-15 e). Then initial nt microstructure frames the shear band (Figure 5-15 d). The grains in the detwinned region (Figure 5-15 b) are on the order of 200-300 nm.

The transition from nt to an ufg microstructure in ntCu20µmLF7 is further analyzed with orientation mapping (OIM) in Figure 5-17. Besides the BF-TEM (virtual bright field, extracted from STEM) micrographs, the crystallographic orientation in x- and y- (both in-plane) and z-direction (out of plane) are visualized, which refer to the [110], [112] and [111] orientation. The mapping covers a region from the nt Cu on the left side approximately 5 µm into the transition zone and the detwinned region on the right, followed by another 1 µm long section with partial nt microstructure. In x- and y-orientation, the color coding gives a good impression of the nt microstructure in the region outside of the shear band on the left hand side of the cross section. The two twin modification alternate with light blue and purple colors in OIM-x and bright red and blue in OIM-y. The orange regions in OIM-x which also refer to the bright green regions in OIM-y result from a failed indexing. In the partially detwinned region in the center of the cross section, roughly 200 nm regions with random but uniform orientations and twins with a reduced density can be found. These regions with different orientations are further referred to as grains. While the orientation does not change in the OIM-x direction, a complete new orientation, indicated by green-yellow colors, arises in the OIM-y direction. This is consistent with the OIM-z direction. Here the regions with the initial nt microstructure on the left hand side have the same orientation. Individual regions, where detwinning occurred and the grains rotated are marked with bluish colors.

The crystallographic orientations of ntCu20µmLF7 are x: ([110], in plane), y: ([112], in plane) and z: ([111], out of plane) what was already shown in Figure 5-1. However, the standard orientation of the OIM is [100] in x and [010] in y direction (both in-plane) and (001) orientation is out-of-plane. The orientations of the OIM are not related to the crystallographic orientations of the sample.

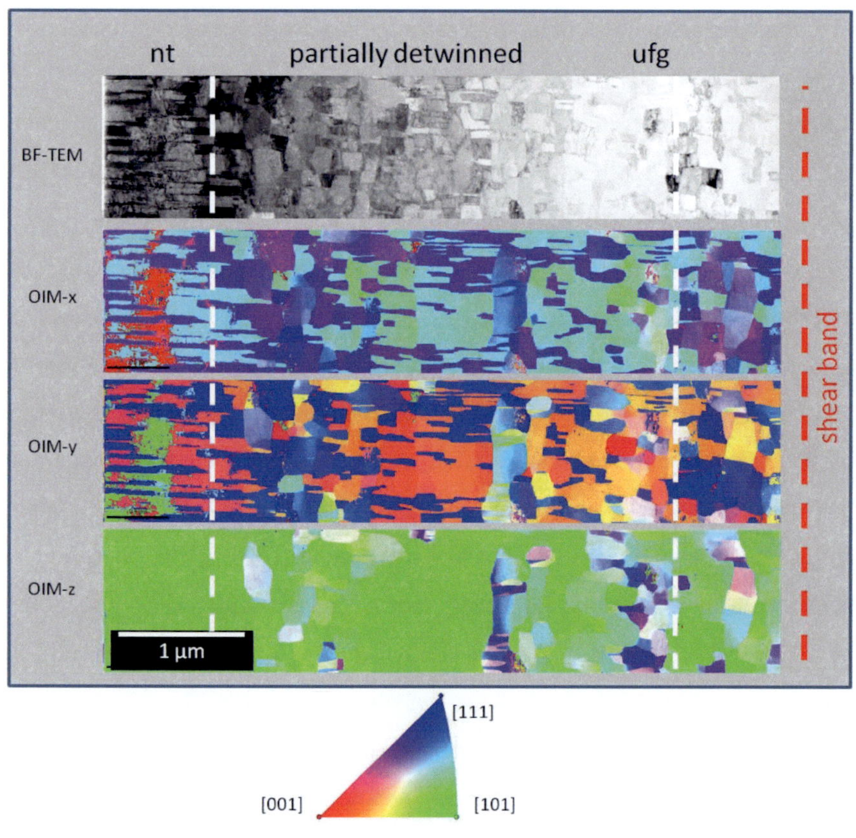

Figure 5-17: (virtual) BF-TEM micrograph and orientation maps (OIM) in x ([100], in plane), y ([010], in plane) and z ([001], out of plane) direction taken from sample ntCu20µmLF7. On the left side of the cross section, the nt microstructure still exists, whereas detwinned and ufg microstructure dominate in the shear band.

The sample ntCu20µmLF8 was stressed until it separated. A TEM-lamella was extracted from the edge of the sample depicted in Figure 5-10 a and Figure 5-18. Surprisingly the microstructure seems to be completely detwinned and the microstructure coarsened, except for a few µm-sized twins in a region of ~ 7 µm away from the edge (Figure 5-18). Close to the very edge of the sample, typical FIB damage is visible [129]. Further away from the edge a domain appears that lost its straight DBs, but is still oriented perpendicular to the surface. Between this domain, which is about 1.5 µm wide and 3 µm high (surrounded by dashed red line in Figure 5-18 b) and the surface remains a region of ultra fine grained (ufg) microstructure (surrounded by blue dashed line in Figure 5-18 b). On the other side of the region, the microstructure seems to be single crystalline again.

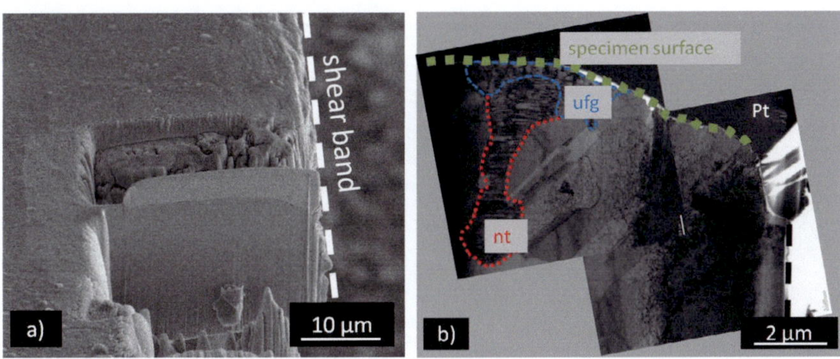

Figure 5-18: SEM micrographs of the ntCu20µmLF8, with the shear band where the sample separated (left). BF-TEM with initial sample surface (green) and the deformed regions - detwinned, still nt (red), and ufg (blue)

The cyclic loading of sample ntCu20µmLF11 was stopped before separation (Figure 5-9 b). In the first 2 µm beside the shear band center only grains but no twins are visible. Between 2 µm and 6 µm away from the shear band (Figure

5-19) a transition region can be observed where the structure density increases gradually and twins have a higher density (Figure 5-20).

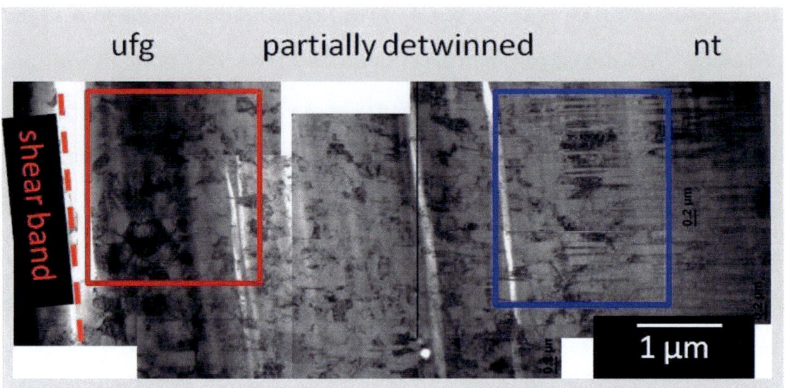

Figure 5-19: stitched BF-TEM micrograph of ntCu20µmLF11 from the region between the shear band (left side) and the still nt region (right side). In between the microstructure is partially detwinned. The regions, where the structure size was analyzed for Figure 5-21 are framed in red near the shear band and blue in the nt region.

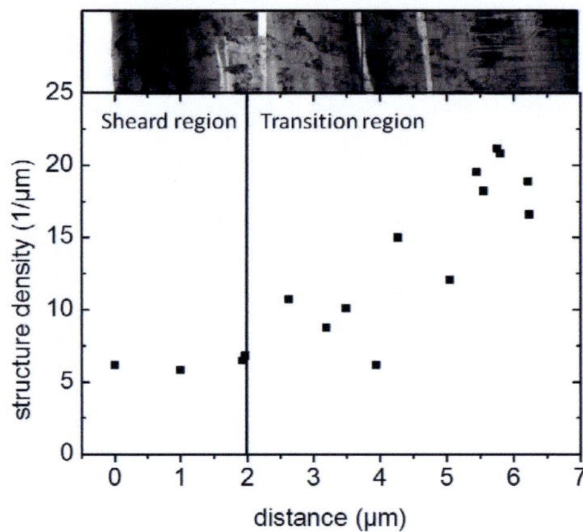

Figure 5-20: Structural density (TBs + GBs) vs. distance from the shear band center. The density stays constant for the first 2 µm until it steadily increases.

The columnar crystalline microstructure in the detwinned region around the shear band did not remain intact due to the cyclic loading and transformed to a more isotropic ufg microstructure.

The evolution of the microstructure in dependence of the distance to the sheared zone is visualized in Figure 5-21 by the cumulative sum distribution of twin spacing or grain size respectively.

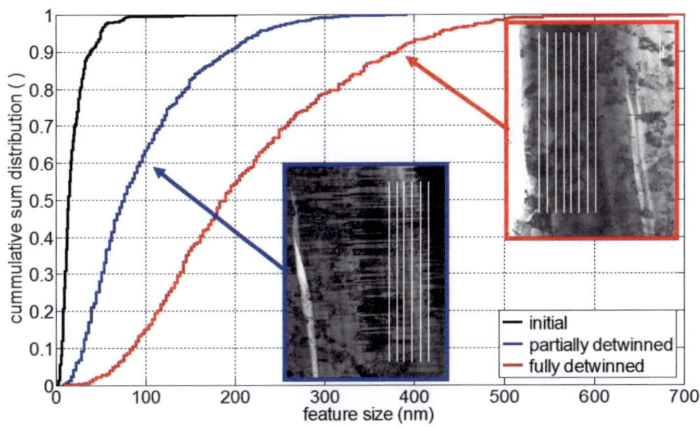

Figure 5-21: Cumulative sum distribution vs. feature size (e.g. twin spacing or grain size) and BF-TEM micrographs of ntCu20µmLF11 in the analyzed regions. The blue framed micrograph shows the partially detwinned and the red framed the detwinned region near the sheared zone. The black curve represents the as deposited nt material (see Figure 5-11 a). The regions that were analyzed with the line intersection method are labeled with the white lines.

The structure density of the detwinned zone near the shear band (red curve) shows grain sizes of up to d_{grain} = 600 nm. In the partially twinned zone 7-8 µm away of the sheared zone, the largest features are on the order of d_{grain} = 400 nm (blue curve). For comparison, the curve of an initially nt material is plotted (black line). This dataset was gathered from the OIM on as-deposited material shown in Figure 5-11 a. The microstructure in the vicinity of the shear band is depicted in Figure 5-22, which shows dislocation debris in the partially nt region with twin spacings of up to ~ 100 nm.

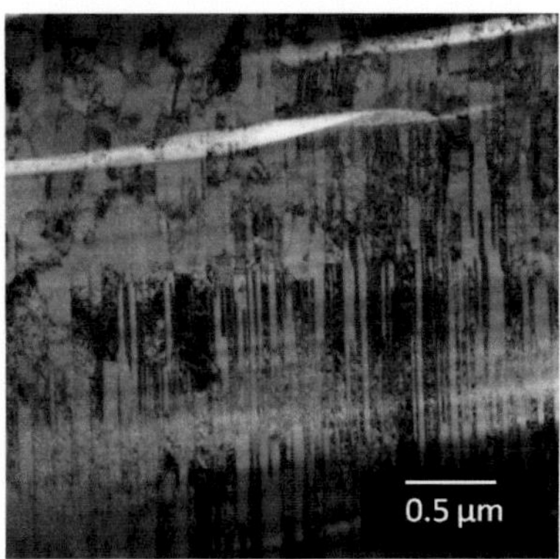

Figure 5-22: BF-TEM micrograph of a partially detwinned region in ntCu20µmLF11 showing dislocation debris between twins.

5.5 Discussion

The presented results are now critically discussed in terms of general aspects of the testing conditions with respect to the sample preparation, dimensions and testing, followed by a section about the mechanical response of the tensile and fatigue tests. Here, the interaction between the mechanical behavior and the nt microstructure are the main interest. Especially the consequences of the detwinning process which might lead to softening will be discussed in detail. The section at the end is focusing on the detwinning mechanisms during mechanical loading.

5.5.1 General aspects

The sample preparation can have a large impact on the test results, especially when the load is applied cyclically. Therefore, e.g. electrochemical polishing was employed to remove residual deformation or processing zones. Another general approach for small samples could be to deposit samples directly in their desired geometry after [171]. If samples have to be cut, the most common processing routes are LASER and EDM cutting processes. In both cases a more or less big heat affected zone of several microns is induced at the cutting plane. These heat affected zones are typically characterized by a coarsened microstructure which can lead to a reduced strength but a good ductility. More detrimental is the brittle surface layer of hydroxides and oxides resulting from the EDM process. The brittle behavior of that surface layer can promote crack formation and propagation from the surface layer into the sample material. This effect is even more severe, when the size of this layer is large compared to the sample thickness. For nc Ni these microcracks lead to a severe reduction in fatigue strength of up to 50 − 75 % [172]. Since the here presented tests were conducted on microsamples cut by an EDM process, it could be argued that the detrimental effects on the properties of the machined surface have led to reduced strength and lifetime.

5.5.2 Tensile testing

One interesting observation during tensile testing is the formation of a shear band in the nt Cu microsamples. The shear band has a very shallow width and leads to a localized shear deformation under an angle of roughly 40-45° relative to the loading axis. Furthermore, at the end of the test, deformation on another shear plane can take place, leading to multiple slip (Figure 5-23). The reason might be the fact that the specimens are gripped at both ends and are not free to move perpendicular to the loading axis. Primary shear is expected to appear at defects while secondary shear is a more geometrically necessary result to adjust to the back stress building up in direction of the tensile axis. Therefore, this leads to the activation of a second slip plane to counteract the rotation. Subsequent shear on multiple intersecting slip planes leads to necking like it is illustrated in Figure 5-23. It cannot be excluded that the shear mode of either single or multiple shear is depending on the loading rate. It appeared in fatigue only at the low frequency but not in all tests.

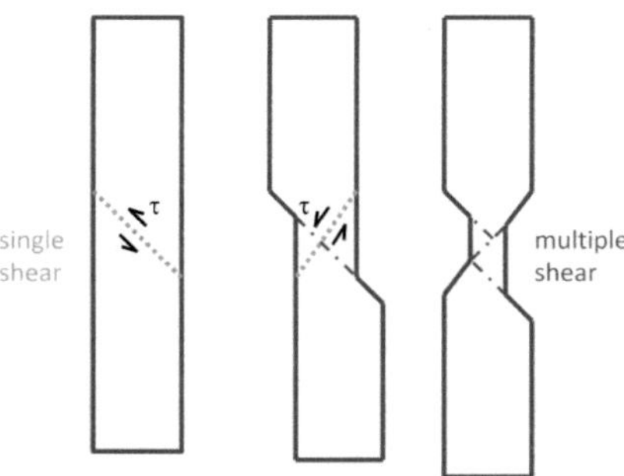

Figure 5-23: Schematic drawing of a loaded sample with single and multiple shear (two slip systems) what is a process that causes necking.

Another interesting observation is the appearance of a yield peak in the stress strain curve during tensile loading. An explanation might be that dislocation based deformation processes cannot keep up with the deformation rate. Similar to polycrystalline nt Cu [162], it can be expected that dislocations can get activated from the ITB. Once they are mobile, they can lead to a fast deformation, and by that lowering the stress. This can result in a yield peak (Figure 5-24). Hodge et al. point out that the sputter process can lead to the low initial dislocation density which in turn can cause the yield peaks [162]. The formation of the shear band in the microsamples correlates with a strong softening from a flow stress of more than 500 MPa to a stress level of roughly 150 MPa (Figure 5-6). This softening can be attributed to a coarsening of the nanotwins. The mechanism behind the microstructural changes will be discussed later.

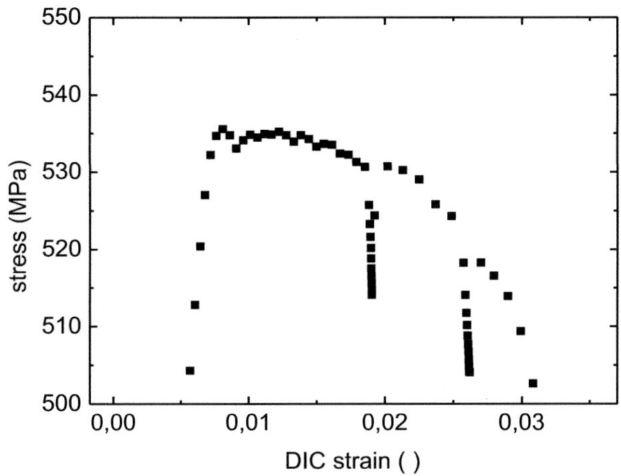

Figure 5-24: Stress vs. strain curve of the tensile test on ntCu20μmT5. At 535 MPa appears a yield peak followed by a decrease in strength.

A Young's modulus of 140 GPa was measured in the loading direction ((111) plane) from the second unloading segment (Figure 5-6). The theoretical modulus along the (111) planes in a [112] direction can be calculated with

$$\frac{1}{E_{112}} = \frac{1}{E_{110}} - 3\left(\frac{1}{E_{110}} - \frac{1}{E_{111}}\right)(\alpha^2\beta^2 + \alpha^2\gamma^2 + \beta^2\gamma^2) \tag{5-2}$$

and

$$\alpha = \cos([112],[100]), \beta = \cos([112],[010]), \gamma$$
$$= \cos([112],[001]) \tag{5-3}$$

from [173] to be on the order of 130 GPa. 67 GPa and 192 GPa are the values for the modulus in [110] and [111] direction [173]. The theoretical value is 8%

lower than the measured value. This is not expected, as the observed porosity at the DBs should reduce the measured value. This deviation is regarded to be in the range of the error of measurement. Due to the small size of the samples deviations of up to 10 % are expected.

5.5.3   Fatigue test

The effect of the mechanical loading on the microstructure in the fatigue tests is similar to the one in the tensile tests. It also leads to a coarsening of the microstructure. The fatigue lifetime of nt Cu are compared to fatigue lifetime of ufg Cu (produced by equal channel angular pressing) from literature in Figure 5-25. The plastic strain controlled tests by Agnew et al. [90] were done at a frequency below $f = 1$ Hz and line up with the fatigue tests at $f = 0.5$ Hz from this work. The tests by Höppel [174] are done with control of plastic strain amplitude at a load ratio of $R = -1$. Load controlled tests on nt Cu at $f = 50$ Hz line up with the ones from Vinogradov [175] who cycled between $f = 10$ and 20 Hz at a load ratio of $R = -1$.

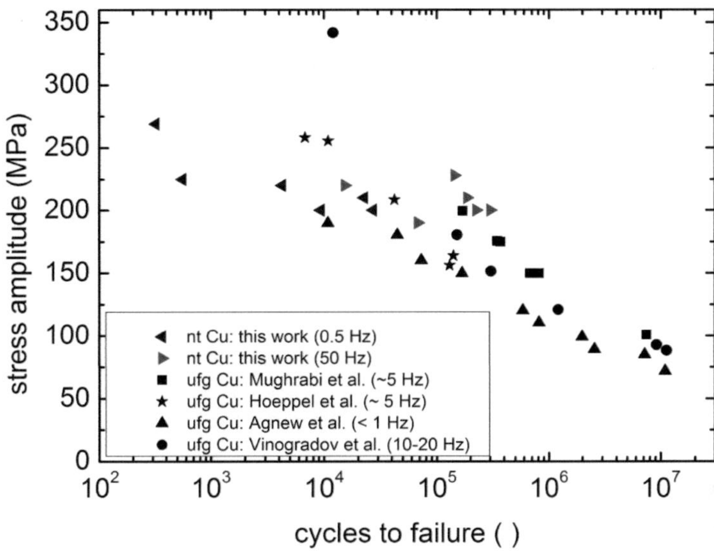

Figure 5-25: SN (Woehler) plot for nt Cu in relation to ufg Cu data from [175], [174], [90] (ufg data is adapted from Mughrabi [83]).

The atmosphere in which fatigue experiments are conducted, e.g. vacuum or oxygen containing, has a strong influence on the life time of cg and ufg Cu. These environmental effects could explain a factor of 10 in fatigue life at a higher loading frequency in cg materials [176]. In these models oxidation of extrusions that formed during fatigue is the classical way of inducing very sharp cracks. While extrusions are pushed out they can oxidize at the surface and when they are pulled into the material again, the oxidized surface becomes the nucleation point for a sharp crack. The nt Cu material seems to follow a different failure mechanism. No dislocation network, PSB and crack formation or crack growth lead to failure, but softening and therefore shear deformation. Therefore, no extrusions or cracks have been observed, strengthening the idea that not the environment affects fatigue life in nt materials but the frequency. This implies, that the frequency effect visible through the difference in fatigue lifetime could be related to a strain rate dependence of the fundamental deformation mechanisms. This effect of increasing strain rate / frequency sensitivity with decreasing twin spacing was observed before [177, 178, 62]. Shen et al. [178] showed in strain rate jump tests that strain rate sensitivity seems to be affected by the interaction of TB and dislocations, and show a stronger sensitivity compared to ufg Cu without nanoscaled twins. Another factor might be the holding segment of the tests on ntCu7 and ntCu8. Both tests were conducted at a frequency below 1 Hz. That means, there was much more time available for diffusion processes.

The reduction of strength and hardness relative to the applied strain rate (loading frequency) can be relevant with respect to the activation volume described by Eq. (2-7) and the dominant mechanism. Nevertheless this context was not further investigated in the light of the activation volume.

No enhanced fatigue lifetime of nt Cu can be detected in comparison to ufg Cu. The strength is higher and the ductility is on a similar level. The coarsening process in nt Cu might need some time to develop, what would increase the fatigue lifetime. Furthermore, the increased tensile strength is expected to increase the fatigue strength [83]. The UTS of nt Cu with σ_{UTS} ~ $500 - 700$ MPa [179] compared to pure ufg Cu σ_{UTS} ~ $300 - 400$ MPa [180] is higher by a factor of two. This is the case if the coarsened region can be regarded as "another" ufg material, which has lost Hall-Petch strengthening during mechanical loading (see Figure 5-6). Such materials tend to strongly localize. Once shear formed at one location, the microstructure degrades and loses its strength. This local deformation has a severe effect as the deformation has to be accommodated by a small volume and a lower fatigue life results. On top of this comes the symmetrical applied load from (R=-1). In contrast to the load ratio R=0.1 of the tests on nt Cu. The more symmetrical load with lower mean stress leads to a less harmful influence of the fatigue. As long as these counteracting influences on fatigue life are balancing out, a similar fatigue lifetime like in ufg Cu could be the result.

The high shear displacement of 55 µm in a narrow band of roughly 15 µm is accommodated by a softening process (Figure 5-9). This behavior is in contrast to the material answer for example in coarser grained Cu, where mutually interference of dislocation and forming of dislocation networks leads to cyclic strain hardening.

In [110] it was observed that very few extrusions formed in materials with grain sizes smaller than 1 µm and it seems clear that in this material with grain sizes between 300 and 600 nm extrusions are not prone to develop. Extrusion formation is expected to be possible in the created ufg structure, even though extrusions were not detected.

5.5.4 Microstructural evolution and deformation

An EBSD mapping was conducted to characterize the orientation relation of neighboring domains. It has to be considered that twins with a thickness of just a few nm might be right underneath the surface. The penetration depth of the electron beam during the EBSD mapping varies with crystal orientation and is on the order of a few nm. Several twins could be involved in the backscattered signal, depending on the twin thickness. As a mixture of signals is difficult to analyze, an indexing rate of roughly 50 % seems to be reasonable. Within the domains, in the nt sputtered material all the twins are elongated perpendicular to the growth and parallel to the loading direction. This means they are "nanostructured" in only one direction. As a consequence the dislocation motion is significantly more constrained perpendicular to the CTB (distance ~ 20 nm) than along the twins (ITB spacing of hundreds of nm). Dimples in the fractured surface of ntCu20µmLF8 were found after rupture as well as lines parallel to the CTB and ITB in ntCu20µmHF11 (Figure 5-10), laterally 10 times bigger than the initial structure. This correlation was already observed by [82]. It is an indication for a coarsened microstructure, but also for deformation mechanisms that are active on a larger scale.

The basic coarsening mechanism during mechanical deformation is detwinning, triggered by partial dislocation motion on (111) CTBs. The direction of dislocation motion is depending on the modification of the twin and the externally applied loading direction [164, 161, 52]. Segments of ITB separate from the domain walls and move in the one or the other direction along the CTB to the next domain wall. The detwinning process is expected to be based on the motion of cells containing paired partial dislocations along the CTB [181]. The paired partials have a height of a multiple of 3 partial dislocations. Thin twins with a short twin spacing might detwin at once by this process. Thicker twins are thinned in a more stepwise manner. The pores that are inherently present in sputtered nt Cu are available as trace marks. In the unloaded state, the pores line up along the domain walls (Figure 5-13 a and b). While in the fatigued state the ITBs are separated from the domain walls and the pores, which remain at their initial position (Figure 5-13 c). This way, the pores can act as fiducial markers, for estimating the relative movement of the ITB. Coarsening of the local twin structure around the shear band can be regarded as a softening mechanism, especially for thin twins of just a few nm (higher velocity [56]). As the TBs have a similar effect on dislocation hindrance as grain boundaries and thus are also an effective Hall-Petch strengthening mechanism [159, 45]. The shear deformation of the nt Cu in the tensile test is very local as described above. Due to the detwinning in the region of the highest stresses, the material loses its strength which was provided by the nanotwins. As they disappear, the local softening leads to a localization of deformation in that region. The consequence is a reduction in strength in combination with a localization of the deformation on a detwinned band across the sample.

Driving force for the twin growth (detwinning) is the reduction of TB area [165]. This trend is in agreement with the micrograph of Figure 5-17, in which detwinning and rotational processes transform the nt into a ufg microstructure. In terms of the orientation of the OIM of ntCu20µmLF7 (Figure 5-17) allows a clear view of the microstructure after mechanical loading. Not just detwinning and a coarsening into ufg microstructure took place but also single grains seemed to rotate, which lead to TB becoming GBs [161, 52]. The rotation of single grains might be accommodated by the huge amount of dislocations which are present after deformation. But the rotation cannot be accommodated by simple partial dislocation motion along the CTB (Figure 5-26 a and b). Even more, dislocation glide systems in the twins and diagonal to the CTB on $(11\bar{1})$ have to become activated (illustrated in Figure 5-26 c). These are hard shear orientations compared to the (111) planes [177]. Additionally to the softening by detwinning the geometric softening due to a reduction in cross sectional area by shear deformation has to be considered, reducing the load-bearing cross section. It is not a global mechanism but happening locally along the shear band, where the highest deformation exists.

Plastic deformation along the loading axis cannot be explained just by partial dislocation motion along the CTBs. Nevertheless, additional to the pure tensile loading condition a shear component on the dislocations on the (111) CTB could arise by experimental inaccuracies. This could result from a slight tilt of the loading axis relative to the CTB. As the partial dislocations on "soft" (111) planes are mobile they do not need high shear stress to move. This partial dislocation activity would create a step at the side of the domains. In sum no plastic deformation would be possible. Instead, for a homogenous deformation of the nt Cu samples more glide systems have to be active, for example on $(11\bar{1})$, where the dislocations move diagonal in the twins (Figure 5-26 c). This dislocation activity on diagonal glide planes was also proposed by K. Lu et al. [63]. In the last step the dislocation activity on other glide planes than the (111) plane can lead to rotational processes and form a ufg microstructure (Figure 5-26 d).

Another process that could lead to this microstructural evolution into an ufg material could be recrystallisation. In the coarsened and softened region of the sheared volume, severe dislocation motion and multiplication is active which consigns significant dislocation debris (see Figure 5-22). This high energy state is a condition of metastable material and is prone to undergo recrystallisation.

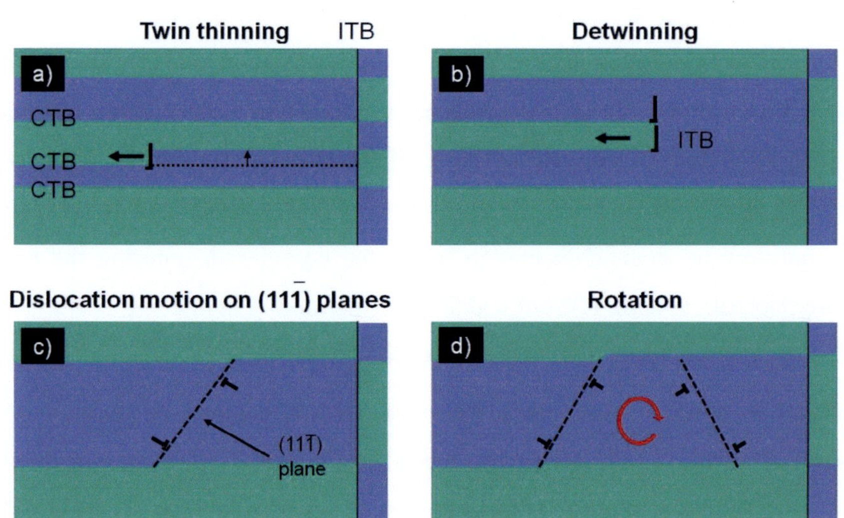

Figure 5-26: (a) Schematic drawing of twinned structure with a paired partials moving along a CTB on a (111) plane leading to twin thinning. (b) Once the twin is thin enough it can get detwinned by the motion of the corresponding ITB. (c) In the next step dislocation motion might be possible on diagonal $(11\bar{1})$ planes if the corresponding twins are thick enough. (d) Finally, dislocation activity leading to rotation or maybe also recrystallisation could form an ultra fine grained microstructure.

Necking not only occurred in-plane by shear on multiple planes to adjust to the tensile axis, but also in z direction. This might be the result of dislocation activity on the diagonal $(11\bar{1})$ planes like proposed above (Figure 5-15 a and f).

Shute et al. [163] presented nanoindentation hardness measurements of nt Cu and pointed out that coarsening could happen, whereas the hardness didn't change. An explanation for this could be the high plastic deformation that leds to detwinning in the highly deformed areas induces high dislocation density and by that a high number of small angle grain boundaries. Similar to what was found in [159], also some Shockley dislocations (partials) were found at the CTB. Those lead to a stress field which is visible in BF-TEM (Figure 5-21) and remains after the test as so called dislocation debris. This dislocation debris is an indication for severe dislocation activities and is visible in Figure 5-22.

Local softening and shear band formation appears to be the dominant deformation mechanism, while crack formation and growth might be hindered by the CTBs [63].

5.6   Summary

1) During fatigue, partials move along (111) CTB planes and lead to subsequent twin thinning. Once the twins are thin enough, whole ITB segments can move and dissolve twins. This detwinning leads to softening.

2) When the twin become thicker, enough dislocations on $(11\bar{1})$ planes can get activated and necking in out-of-plane direction occurs.

3) This coarsening and rotational process (recrystallisation might be involved) transform the initial nt Cu into ufg Cu.

4) Failure occurs by shear along the softened regions, opposite to cracking in ufg and cg Cu

5) Nt Cu and ufg Cu have similar fatigue life at comparable frequencies

5.7 Acknowledgement

Thanks are given to Dipl.-Ing. Stefan Slaby for his help with the sample preparation as well as Dipl.-Phys. Paul Vince for the AFM mapping, Dipl.-Ing. Simone Schendel for the EBSD mapping and Dipl.-Ing. Heino Besser for the sample machining with the LASER. Finally Prof. Gumbsch is thanked for helpful discussions.

6 *In situ* TEM mechanical testing of nanotwinned Cu

6.1 Abstract

In situ TEM tensile and fatigue tests were conducted on nanotwinned (nt) Cu samples. The sputtered nt Cu was single crystalline, but contained domains with sub micrometer diameters of stacked twins grown in the [111] direction. In contrast to electrodeposited nt Cu, the coherent twin boundaries (CTBs) were aligned perpendicular to the growth direction. The twins are only nanostructured (ns) along the in-plane orientation. TEM tensile testing samples were extracted perpendicular to the twin boundary (TB) plane and then fixed to a push-to-pull (PTP) device. These samples were tested under tensile monotonic and cyclic loading. The test setup allowed us to observe the microstructural changes *in situ*, as partial dislocation motion led to thinning of twins. This might be the first step of the detwinning process. Furthermore, the setup allows us to test small material volumes on the order of 10^{-11} mm^{-3}. This size regime is not accessible with conventional setups. Nevertheless, tests on those samples are of interest while dominant deformation mechanisms and properties are different than in conventionally sized samples. In the tensile tests, an ultimate tensile strength (UTS) of up to 2.5 GPa was measured. A new method for extracting and preparing tensile testing samples of any material and orientation will be discussed.

6.2 Literature Review

Nanotwinned (nt) metals show enhanced tensile strength and ductility with decreasing twin size [43]. They gain their improved mechanical performance, such as strength of more than 1 GPa from the TBs, which act as effective dislocation sources and barriers at the same time [43, 63]. It was observed that during fatigue this nt structure can dissolve and loose its ability to promote the material's strength. This process is called detwinning [165, 164, 182] and seems to be triggered by partial dislocation activity along the coherent twin boundaries (CTBs).

Full dislocations were observed within twins down to a twin spacing, λ, of 30 nm [55]. Below that size, the motion of partial dislocation is expected to be the dominant deformation mechanism [183]. The consequence is a competition between the stress needed for the nucleation and motion of a full dislocation at higher twin thicknesses and the motion of Shockley partials and twin boundary migration (TBM) at smaller twin sizes.

Wang et al. showed that at the intersection of $\Sigma 3$ (111) (CTB) and incoherent twin boundaries (ITB), partials with $\left|\vec{b}\right|$ = [$\bar{1}\,\bar{1}\,2$]/6 are emitted [165]. These partials can form ledges at the CTB, which act as stress concentrators and sources for additional dislocation nucleation and emission [56]. Steps of three (111) planes or multiples of them [161] move along the CTB when a load is applied, and appear as ITBs. The direction of the TBM depends on the applied load and configuration of twin orientations. This is illustrated in Figure 6-1 where partials formed at the intersection of the domain boundary (DB) and the CTB and move as ledges (L1, L2 and L3) along the CTB under an applied load. As a consequence, the twin thickness changes at each ledge.

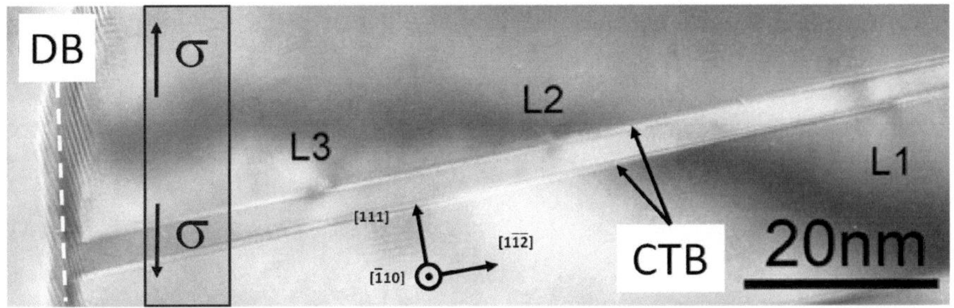

Figure 6-1: Bright field transmission electron microscopy (BF-TEM) micrograph of nt Cu with a DB on the left-hand side and two CTBs. When the sample is under a load, Shockley partials (L1, L2, L3) are emitted from the intersection of the DB and the CTB and move along the CTB, depending on the stress state (adapted from [56]).

Similar to nanocrystalline (nc) materials, the yield strength σ_y in nt materials increases with decreasing twin spacing λ, following a Hall-Petch like relation [43]:

$$\sigma_y \sim \sqrt{\frac{1}{\lambda}} \qquad (6\text{-}1)$$

As full dislocation activity becomes energetically unfavorable in a material with a twin spacing below 30 nm, Hall-Petch scaling fails to describe the material's strengthening [184]. The reason is that the distance between the TBs, which are emitting the dislocation on one side and where the dislocations would pile up on the other, are too small to bear any dislocations. Alternatively, the partial dislocation extension model is used to describe the strengthening in nanostructured materials [23, 22, 40] and [185] follow the relation:

$$\sigma_y \sim \frac{1}{\lambda} \qquad (6\text{-}2)$$

In this model, the increase in strength is related to the generated stacking fault (SF), when leading and trailing partials are emitted from the TB. The inverse Hall-Petch regime is reached for $\lambda < 15$ nm, where the material softens or reaches a plateau with decreasing λ. Models for a kinetic and energetic description of partial dislocation mediated correlation in fcc metals are given by [182]:

$$\sigma_y \sim \ln(\lambda) \tag{6-3}$$

and

$$\sigma_y \sim \lambda \tag{6-4}$$

The influence of the domain size is not considered here, as it is one order of magnitude larger than the twin spacing. For nt Cu, not only increasing strength and increasing ductility were observed with decreasing twin spacing, but also increase in toughness, fracture initiation resistance and damage tolerance during stable fatigue crack growth [54]. Twins can act as obstacles to crack extension, thereby impeding crack growth [56] and increasing crack path tortuosity in a zig-zag manner. The edges are roughly between the atomic planes in $[\bar{1}10]$ and [112] direction [55]. Therefore, the nt structure has a major influence on the fracture toughness. An increase of K_{IC} from ufg Cu to nt Cu was observed in [54] (Table 6-3).

As the structural dimensions of metallic materials have the trend of becoming smaller and smaller, the testing methods capable of examining those structure sizes have to be suitable. Possible techniques are bulge testing [186, 187] and nanoindentation [188]. Both techniques have the disadvantage of a complicated stress state and therefore a comparison to results from standard tensile tests becomes difficult.

Due to the complicated deposition method of nt materials by sputtering or electrodeposition, they are basically only available in the shape of thin films of less than 20 μm. *In situ* SEM/TEM [94, 96] tensile testing of thin films (or thin whiskers) is a more advanced test, with a defined application of load while observing microstructural evolution *in situ* during the test. The electrochemical deposited films form randomly oriented twins while the sputter deposited film can form parallel twins [63] and are available in higher purity. Another feature of the sputter process is that the films can be deposited as a single crystal. The sputter deposition of nt Cu is described by Zhang et al. [110]. The authors deposited films with a thickness of approximately 20 μm and showed that it consisted of ultra fine sized columnar domains. These domains are columnar stacks of twins in the [111] growth direction. The twins were reported to be oriented perpendicular to the growth direction with a mean twin spacing of ~ 25 nm (Figure 6-2).

One way of preparing samples for in-situ TEM tensile testing is to sputter or electrodeposit a thin electron transparent film either on a substrate that can be dissolved (for example NaCl) or on a substrate, where the film can easily be separated (for example mica). In case of a dissolvable substrate, the films float on the surface after the thin film and the substrate is carefully inserted into water at a flat angle. After this separation the film is captured with a testing structure. To achieve a fully quantitative tensile test rather than a qualitative straining experiment, a "push-to-pull" (PTP) device that transforms the compressive load of a picoindenter into tensile load on the sample can be used [99]. A key point of this thin film approach is that materials can only be tested along their in-plane orientation, so perpendicular to the growth direction.

In order to test materials along their other orientations, a novel method to extract and fabricate nanosamples by FIB for *in situ* TEM tensile testing was developed in this study. Therefore, a new methodology will be presented, which allows extracting TEM tensile samples from bulk materials for *in situ* testing. The mechanical behavior, crack propagation and microstructural evolution will be presented and discussed in the following sections.

6.3 Experimental

In this section the deposition of the nt Cu films, the *in situ* TEM sample preparation and the *in situ* testing procedure will be described in detail.

6.3.1 Extraction of the TEM tensile testing samples

The nt Cu films were sputter deposited at the University of Texas in the group of X. Zhang by high rate magnetron sputtering at a base pressure of less than $5*10^{-8}$ torr on a (100) oriented Si substrate. The sputter deposition of singe crystal nt Cu has also been described by Anderoglu et al. [45]. The tested films were approximately 20 µm thick and consist of ultra fine sized domains of columnar stacks of twins with diameters between 200 and 600 nm. Since the twins are oriented perpendicular to the growth direction (Figure 6-2), the twins are not visible from top view. To be able to observe the deformation in the material during *in situ* loading the samples should be extracted with their tensile axis perpendicular to the growth direction. The sample orientation was chosen in a way to allow for a cross sectional view of the twins in the TEM.

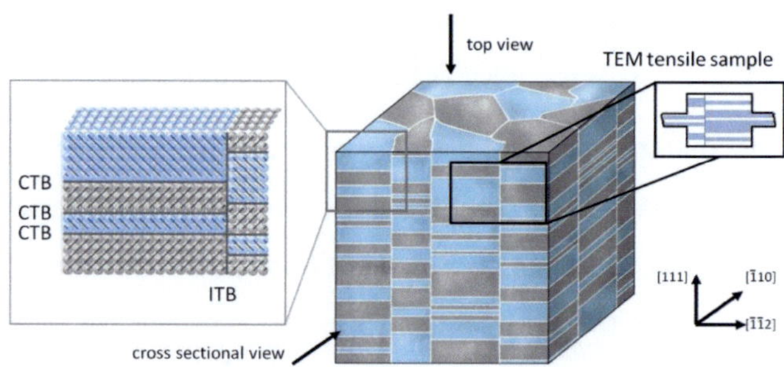

Figure 6-2: Schematic drawing of the investigated nt Cu with twins parallel to the surface and therefore not visible in direct top-down-observation. The domains are directionally solidified along the growth direction. The geometry and orientation of the testing lamellas are also depicted.

For *in situ* TEM tensile testing, samples with a final thickness of less than ~ 100 nm are desirable to ensure electron transparency and a clear view of the TBs and the damage formation. The stepwise sample fabrication is shown in Figure 6-3. First, a cross section with a thickness of 1.5 μm, a length of 55 μm and a depth of 15 μm is cut using a Focused Ion Beam system (FEI Nova nanolab 200 dual beam, Netherlands) at 30 kV and a beam current of 20 nA. The cross section is then thinned to ~ 1 μm thickness at 30 kV and a beam current of 1-3 nA. In a second step, the sample is then tilted to 52° relative to the ion beam to cut the final shape at 30 kV and a beam current of 3 nA (see Figure 6-3 a). After this step, the structure has a 8 μm long straight section between the two pulling arms and is still fixed on both ends to the base material. Afterwards this structure is glued to a micromanipulator needle, released on both ends by a FIB cut, transferred to an Omniprobe lift-out-grid and fixed with Pt from the gas injection system (Figure 6-3 b). For the last steps of the extraction process, the sample is tilted back and a 5 μm long part of the straight section is thinned at 30 kV and 0.1 nA to a thickness of roughly 1 μm. To ensure a constant thickness throughout this sample section, the thinning step to 200 nm is carried out at an angle of + and - 1.5°. Finally the sample section is thinned to electron transparency at 5 kV and 70-10 pA and angle of + and - 5° (Figure 6-3 c). This also is meant to remove the damaged layer created in the cutting process. While only the center section of the sample is thinned to electron transparency, the top and bottom remains thicker to keep the sample mechanically stable for the following transfer step. The sample is glued to the PTP device at the end of the arms using the gas injection system of the FIB tool. To prevent an accidental Pt-deposition in the gage section the arms should be at least ~ 10 μm long.

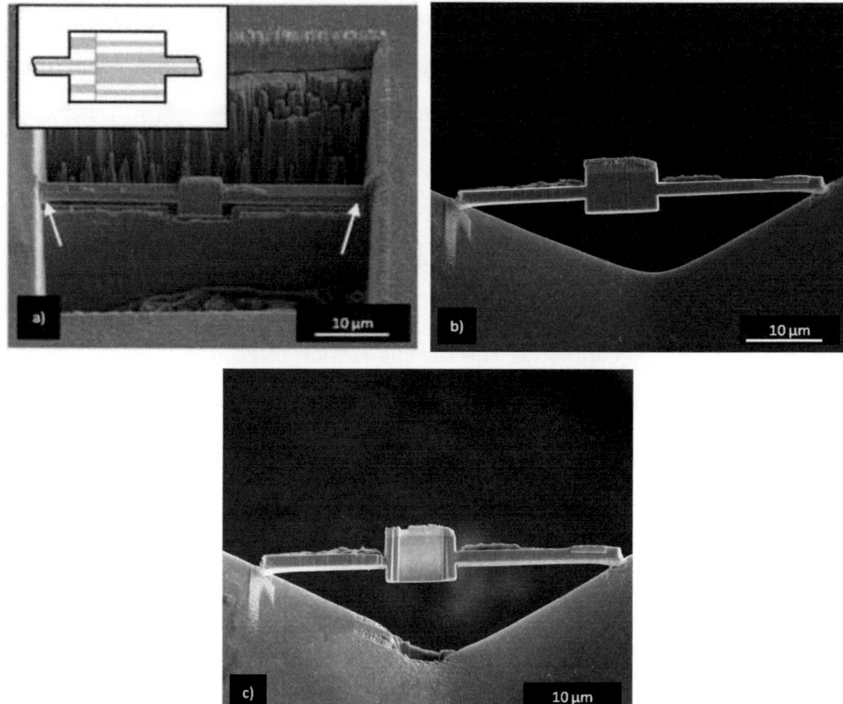

Figure 6-3: (a) SEM-micrograph of the shaped tensile sample in the deposited film. The white arrows indicate, where the TEM tensile sample is finally released. (b) the sample is extracted and glued to the lift-out-grid. (c) the mid section is thinned to electron transparency at 30 kV, 0.1 nA and 1.5° tilt and finished at 5 kV, 10 pA and +/-5° tilt.

6.3.2   Transfer of the TEM tensile sample to the PTP device

In order to transfer the sample on the PTP device, the lift-out-grid with the sample must be dismounted from the FIB and tilted around the main axis of the sample by 90°, in order to align it with the PTP device (Figure 6-4 a). Back in the FIB the tensile sample can be transferred to the PTP with the correct horizontal orientation where the thinned section is situated between the crossheads (Figure 6-4 b). It is important to note that the axis of the sample and the lift-out-grid should be aligned with great care as this defines how horizontal the sample can be placed on the PTP.

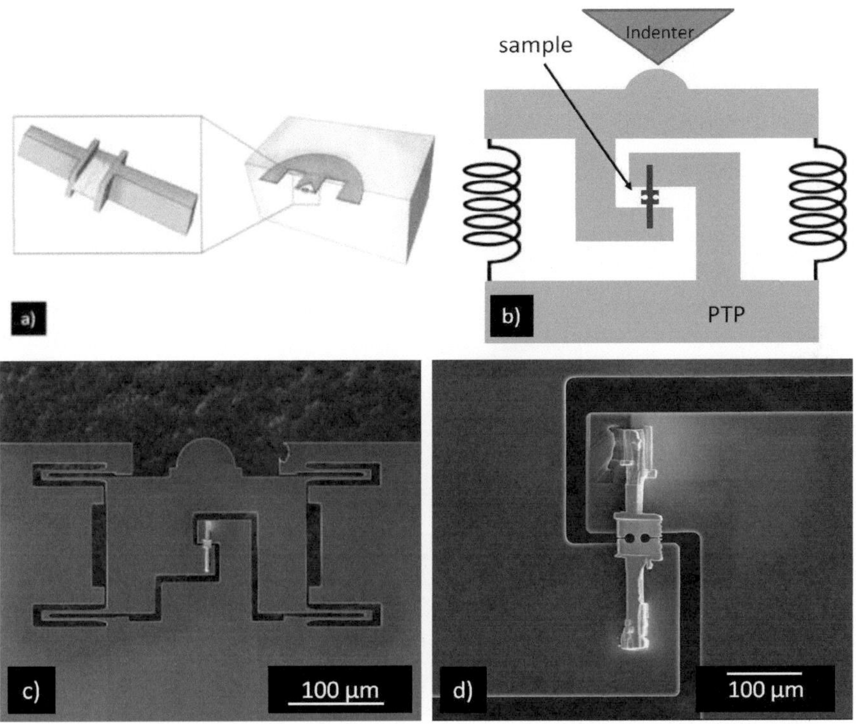

Figure 6-4: (a) Schematic drawing of the sample still fixed to the lift out grid. (b) schematic drawing of the principle of a PTP. (c) SEM micrograph of the tensile lamella transferred from the lift out grid onto the testing geometry of the PTP device. (d) SEM close-up view of the sample fixed to the PTP.

Once each arm of the sample is fixed to the PTP respective cross head, the thinned section can finally be shaped. For dog bone-shaped (Figure 6-5 a) and hour glass shaped samples (Figure 6-5 b), straight and annular milling patterns at 30 kV and 10 pA are used to minimize the damage layer at the edges of the samples. Some of the samples were intentionally notched during this step, in order to have a defined "site of interest" where the sample is most likely going to fail and observations at higher magnification are possible (Figure 6-5 c).

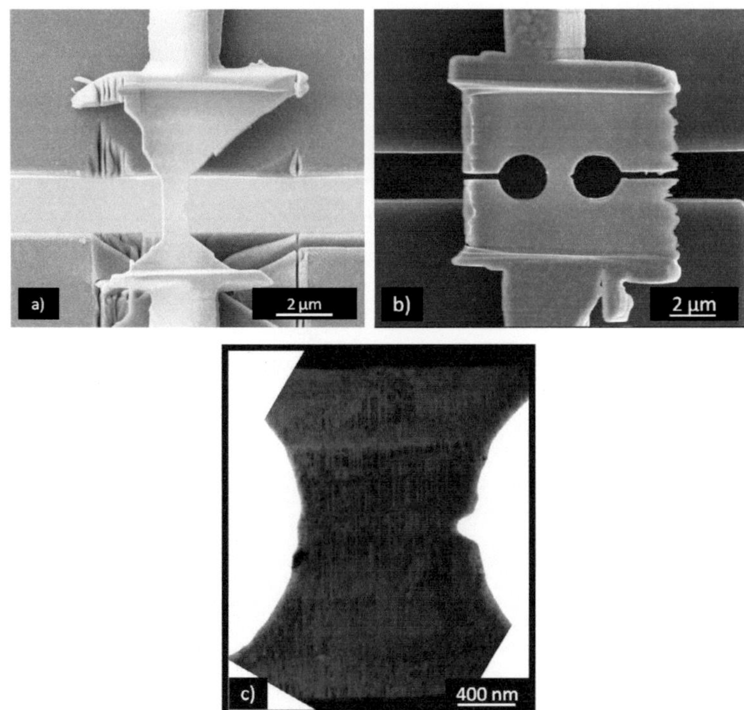

Figure 6-5: (a and b) SEM-micrographs of the two types of notched TEM tensile samples with a straight dog bone shaped geometry and an hour glass shaped geometry. (c) The STEM micrograph visualizes the notch in the hour-glass shaped sample with a gage width of roughly 1 μm. In this illustration, the ITBs are oriented horizontally and the CTBs are vertically.

6.3.3 *In situ* tensile testing

All tensile and cyclic loading tests were carried out in a displacement control manner. The stress was somewhat difficult to estimate as the samples did not have a uniform sample thickness, due to a thin layer of redeposition and the curtaining effect. Nevertheless, the thickness of the samples was measured at different spots in the gage section with electron energy loss spectroscopy (EELS) [189].

Furthermore, the displacement control offers an important advantage in case of fracture, as crack propagation is more stable. Therefore, it is easier to observe microstructural changes during crack growth. First cyclic tests were performed to a maximum displacement of $30 - 50$ nm without failure, after which the maximum displacement was increased to about 120 nm and cycled until fracture occurred. After failure, the thickness was measured using EELS in the vicinity of the crack to examine the amount of necking. The nominal stiffness of the PTP is in the range of 350 N/m. Nevertheless, each PTP was individually loaded after sample failure to determine the individual stiffness.

Strain was measured by digital image correlation (DIC) [138] from the recorded videos, using the edge of the PTP crosshead and, if possible, across the sample (Figure 6-6 a).

For comparison, single crystalline Cu (grain size of several µm) thin films were tested in the same manner.

The following label convention is used: The name of each sample consists of the base material (nt or cg Cu for the single crystalline sample), second the film thickness in nm, third the testing mode (T for tensile and C for cycling) and last an ongoing number.

6.4   Results

6.4.1   Mechanical properties

Dog bone and hour-glass shaped nanosamples with a thickness between 46 nm and 127 nm were tested *in situ* in the TEM. The measured ultimate tensile strengths (UTS) are listed in Table 6-1 with an error estimated by the Gaussian error propagation (see chapter 4) and also the yield strength σ_y. According to the thickness deviation from the EELS measurements the estimated error in width was set to 20-25 nm (depending on the sample geometry). The error for the load cell was estimated to be 10 µN. For cyclic tests the listed UTS corresponds to that reached in the last cycle.

Table 6-1: Sample dimensions of the tested nt and single crystalline Cu and resulting ultimate tensile strength (UTS) with errors. Sample T1, C3 andT5 were notched. All samples had a width of 1000 nm except ntCu83nmT1 had 859 nm and ntCu100nmC2 had 1090 nm and ntCu55nmC3 had 884 nm.

sample	cycles	h (nm)	disp. rate (nm/s)	UTS (MPa)	σ_y (MPa)
ntCu83nmT1	1/4	83 +/- 16	8	1964 +/- 205	1470
ntCu100nmC2	4	100 +/-15	5	2651 +/- 170	1700
ntCu55nmC3	4	55 +/- 17	3	1604 +/- 180	1440
ntCu127nmT5	1/4	127 +/- 10	-	-	-
cgCu64nmC1	9	64 +/- 3	4.3	1797 +/- 91	1523
cgCu67nmC2	3	67 +/- 14	3.5	985 +/- 128	909

The measured UTS ranges from σ_{UTS} = 1600 MPa to σ_{UTS} = 2650 MPa for the nt Cu samples. The exceptionally low strength of σ_{UTS} = 670 MPa for the sample ntCu46nmC4 was a result of a damaged PTP device which showed fracture prior to the test. The single crystalline samples showed an ultimate strength of σ_{UTS} = 990 MPa and σ_{UTS} = 1800 MPa.

The strain of ntCu100C2 was analyzed with DIC between each end of the gage section and the edge of the PTP (see Figure 6-6 a). The stress vs. strain curve is shown in Figure 6-6 b. Final fracture occurred in the region of the left side of the gage section. Total strain was in the range of $\varepsilon \approx 2.5$ %, with a yield point at $\sigma_Y = 1.5$ GPa and roughly $\varepsilon \approx 1$ % elastic strain. The Young's modulus was measured to be $E \approx 168$ GPa $\pm$ 10 %. In Figure 6-6 b the red stress-strain curve is determined from the displacement related to the initial gage length.

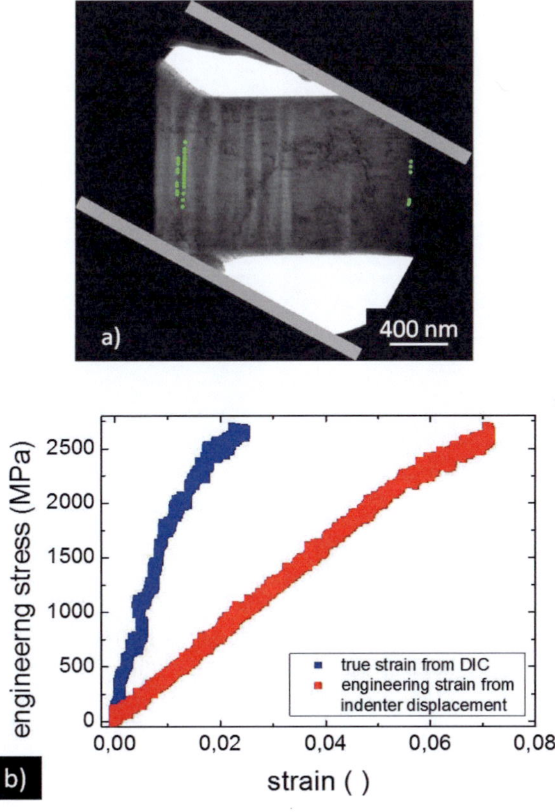

Figure 6-6: (a) BF-TEM micrograph of ntCu100nmC2 with fiducial markers used for strain measurement with DIC, and (b) a stress strain curve of the corresponding specimen.

Sample ntCu100nmC2 formed a crack at the notch during the 4th cycle. During that cycle, the crack grew roughly 25 nm and stopped at the first CTB. The further increasing stress led to the fracture of the CTB, which led to final fatal failure. The crack propagation is shown in Figure 6-7 in the micrographs a-f, which were extracted from the video that was captured during that cycle.

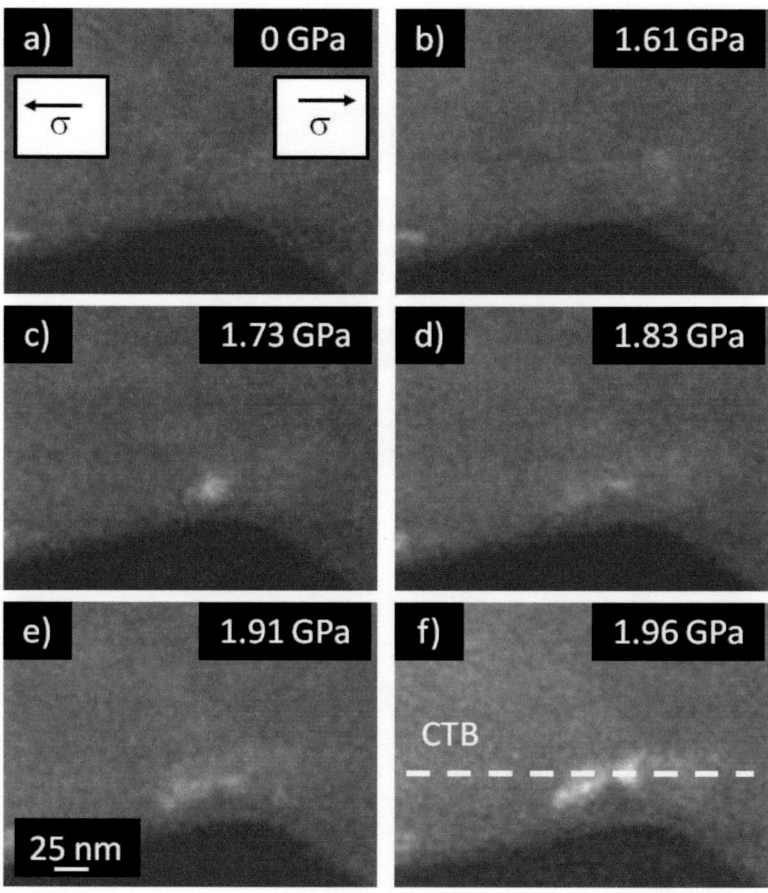

Figure 6-7: Series of DF-TEM micrographs that are captured during the last cycle of loading in ntCu83nmT1. The loading axis is horizontal as indicated by the arrows. In (a) the notch is unloaded and forms in (b) and (c), propagates between (d) and (e) and stops in (f) at a CTB right before (~ 0.03 s) final failure.

Taking into account the crack formation and propagation until it reaches the first CTB, the stress intensity reached $K_Q = 6.7 \pm 0.8\ MPa\sqrt{m}$ was estimated by an approach described in [54]. The values used are listed in Table 6-2.

Table 6-2: List of parameters used and determined stress intensity value for a crack length of 25 nm.

	$a \sim 25 \pm 5$ nm
$dW\ (10^{-11}\ Nm)$	7.2
$dE_{el}\ (10^{-11}\ Nm)$	5.9
$dE\ (10^{-11}\ Nm)$	1.3
$dA\ (nm^2)$	5000 ± 500
$G_{IC}\ (N/m)$	266 ± 20
$E\ (GPa)$	168 ± 17
$K_Q\ (MPa*m^{1/2})$	6.7 ± 0.8

6.4.2 As deposited microstructure and its evolution

Figure 6-8 shows a BF-TEM micrograph with as deposited ultra fine sized domains and embedded nanoscaled twins. The domains are grown columnar in [111] direction, which is marked with the white arrow. The TB spacing is on the order of 25 nm.

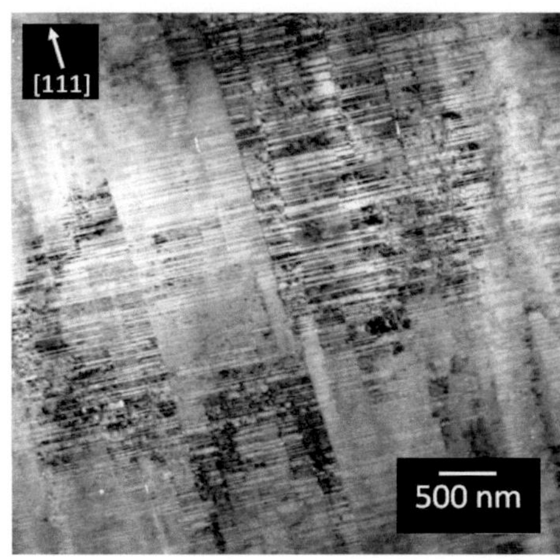

Figure 6-8: BF-TEM micrograph of the as deposited microstructure of magnetron sputtered Cu containing nanoscaled twins.

The test of the ntCu127nmC5 was performed at an unconventionally high strain rate. Therefore no data for load and displacement is available. Fracture occurred in a zig-zag manner under ± 54° with respect to the CTB. Crack deflection occurred at the CTB (Figure 6-9 b). This results in a longer fracture path and crack tortuosity than in the tests done at a lower displacement rate of 3.5 and 4.5 nm/s (Figure 6-9 a). Observations indicate that this sample sheared off in direction to the normal of the gage section (Figure 6-9 b). On the fracture surface, which was created in the vacuum of the TEM, no surface layer such as oxide and contamination is visible as can be seen on the left side of the sample (Figure 6-9 b).

Figure 6-9: BF-TEM micrograph of (a) the nt Cu tested at low strain rate. (b) In comparison zig-zag fractured surface in ntCu127nmT5. Clearly visible is the fractural shape under 54° at high strain rate.

In some of the twins near the fracture surface, as well as on the sheared and therefore oxide free part of the sample, dislocation ledges could be identified. The thinnest twins exhibiting such ledges are down to ~ 10 nm. Those features are labeled with A-D in Figure 6-10 a. Feature E in Figure 6-10 a. is shown more clearly in Figure 6-10 b. In this picture, the diffraction conditions are slightly changed and feature E can be identified as a straight ITB

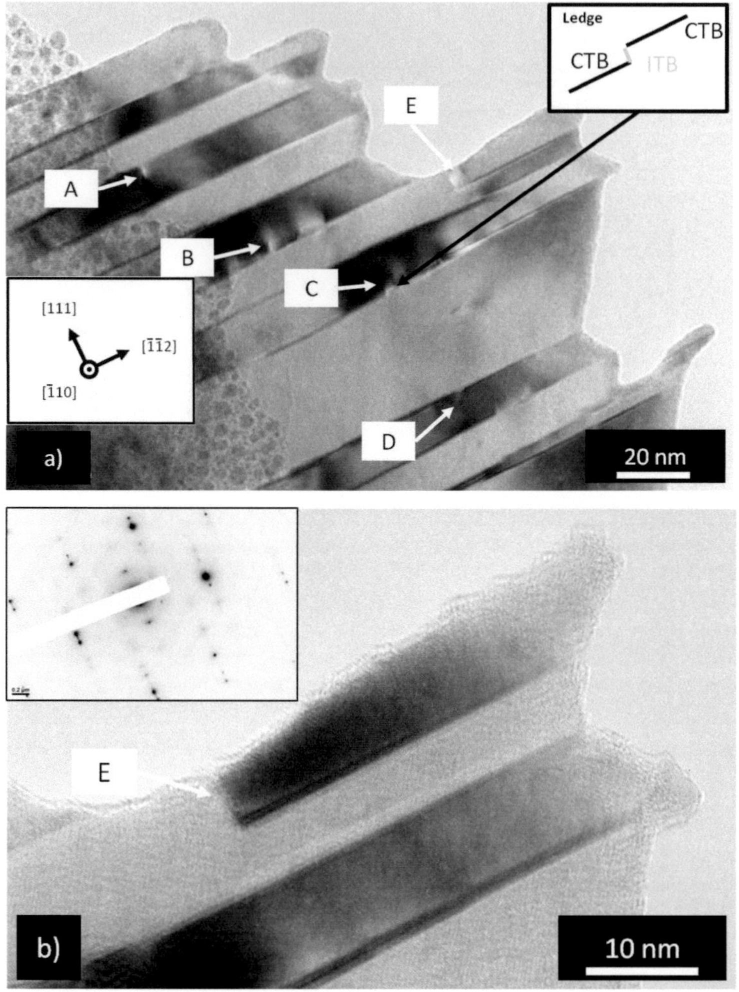

Figure 6-10: (a) BF-TEM micrograph of the sheared surface after the high strain rate test. (b) The upper dark twin next to feature E is separated by an ITB and by a CTB of the lower and brighter twin at different diffraction conditions and a diffraction pattern of ntCu127nmT5, clearly visible the double reflex resulting from the twinned structure.

Those twins which contain ledges along their CTBs have a coned shape with decreasing thickness towards the fracture surface. Single ledges in the shape of paired partials with a height of ~ 1 nm appeared even on ~ 6 nm thin twins

(Figure 6-11 a). The arrow in Figure 6-11 b points out a region which is detwinned at a length of ~40 nm by motion of a section of an ITB. Nevertheless, no global detwinning was observed.

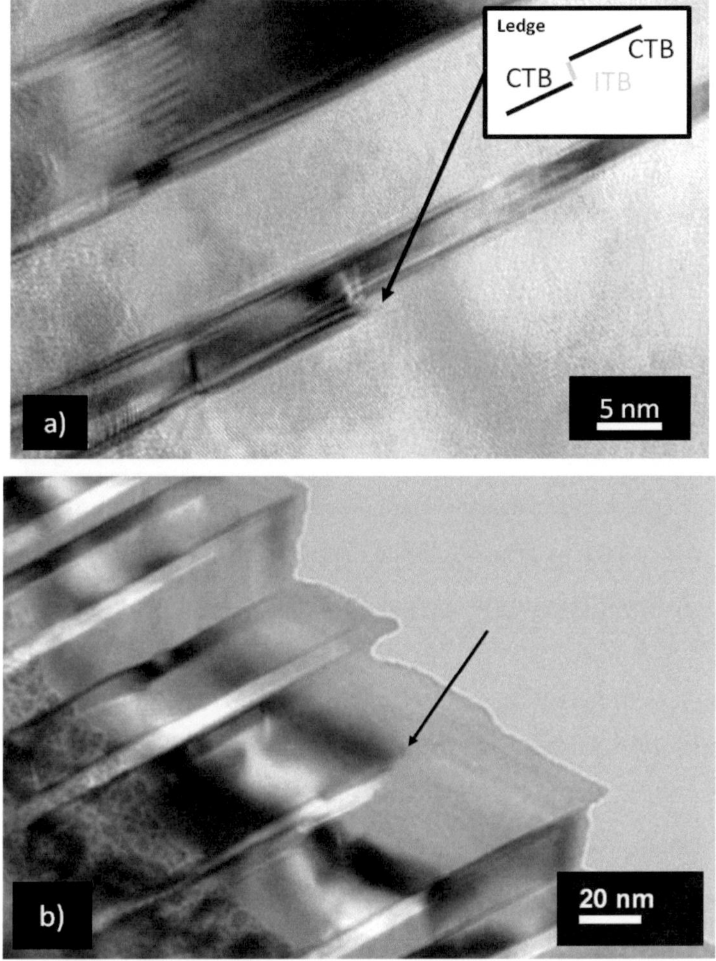

Figure 6-11: (a) BF-TEM micrograph of a ledge on a CTB in ntCu127nmT5, and in (b) a section, where a thin twin is completely detwinned next to the fracture surface.

6.5 Discussion

In the following section, the previously mentioned results are critically discussed in terms of the experimental methods and which challenges or errors can arise. The following section will discuss the mechanical properties, microstructural stability and detwinning by partial dislocation motion and finally the observed fractography.

6.5.1 Experimental methods and data evaluation

An important point that has to be considered in *in situ* TEM testing is electron beam induced damage. Radiation damage by knock-on collision has a threshold value for Cu U_{th} = 400 keV [190]. Since the used TEM had an acceleration voltage of U = 200 kV, knock-on collision damage was not expected to be an issue. Coarsening of the twins and domains at RT are not expected due to the low interface energy [129]. Nevertheless e-beam induced specimen heating of the sample during TEM observation could increase the thermally available energy. Cu has a high thermal conductivity of 400 (W/K m), and therefore at a uniform illumination with a field view of R= 20 µm at a current of j= 10^{-2} A cm^{-2} no changes in microstructure are expected [191]. Ga$^+$ ion damage in Cu by ionization in the FIB cutting process has been observed and manifests in a roughly 10-20 nm thick amorphous layer at the edge of the sample and can lead to embrittlement.

A contamination layer is clearly visible at the sheared surface of Figure 6-9 b. This can be attributed to an unwanted deposition of Pt during the sample fixing procedure or to redeposition during FIB cutting. An indication for this is that the surface region of Figure 6-11 which was created during the experiment due to fracture inside the TEM, appears to have a better contrast. This layer of contamination could influence the mechanical behavior, if cracks in that layer can lead to crack nucleation in the sample. Furthermore, that layer could influence the thickness measurement by EELS. In case of redeposited Cu, it is not expected to have an influence on the measured thickness while in case of Pt, the thickness of the sample could be overestimated.

Furthermore, the load cell can show oscillations, which are visible in the stress-strain curve in Figure 6-6. This was attributed as an artifact of the electronics and data acquisition. From the experience so far it does not seem to be related to an oscillation of the picoindenter transducer and therefore does not correspond to undesired load oscillations on the sample.

As the *in situ* TEM holder allows no double-tilt, the tested samples cannot be investigated in HRTEM mode. In order to conduct HRTEM investigations the sample needs to be extracted from the PTP device and mounted onto a lift out grid for examination in a conventional double-tilt TEM sample holder. Since the sample is plastically deformed during the test, the two parts of the fractured sample are elongated and can get into compression during unloading. Here, a design upgrade of the PTP device could help if a clamping mechanism could be added.

The characteristic contrast changes in the TEM e.g. due to bending contours or dislocations can make it difficult to use DIC techniques to measure the strain inside the gage of the sample during loading. Even though, DIC strain measurement is not straight-forward it can still be applied to track contrast edges that result from thickness contrast instead of diffraction contrast. E.g. the ntCu100nmC2 sample offered contrast features, independent of loading conditions and positions resulting in a sufficiently accurate strain analysis. A solution could be to record the video in STEM mode, since only inelastic scattered electrons contribute to the image function. As a result, e.g. bending contours have only little effect on the contrast and therefore DIC could be more easily applied.

In Figure 6-6 b, the stress vs. strain curves for the strain determined from DIC (blue) and from the indenter tip displacement (red) are plotted. The three times higher strain in the red curve might be explained by the large distance between the fixing points of the sample on the PTP (see Figure 6-4 b) and the stiffness of the overall setup.

6.5.2 Mechanical properties and microstructural stability

Lu et al. [43] described how the twin boundaries can be effective barriers to dislocation motion. As the domain size is roughly ten times larger than the lamella spacing, the twin spacing λ seems to be the major contributor for the mechanical strength. The ultra fine domain size could also lead to a Hall-Petch strengthening but its impact would be expected to be smaller [192]. Therefore, it seems plausible that the dislocation activity is mostly concentrated on the (111) glide plane parallel to the CTB within the domains and not one of the glide planes inclined relative to the CTB. From dislocation theory the stress necessary to nucleate a dislocation in nt Cu with a twin spacing of 30 nm can be estimated to be on the order of 700 MPa [55]. As is shown in Figure 6-10 a, nanotwins containing detwinning partials were found down to a λ of ~ 11 nm. With a linear-inverse correlation (Eq. (6-2)), this implies a yield stress of $\sigma_y =$ 2.1 GPa. Thus, the whole strengthening effect seems to be a mixture of the partial extension model and the twinning dislocation model due to the broad twin spacing distribution [23, 40]. Nevertheless, due to the unique sample preparation technique, the sample contains a lot of defects that can act as stress concentrators, for example notches at the edges or free surfaces. Even higher strength could be expected in samples with smooth surfaces and an optimized shape.

The measured Young's modulus of 168 GPa is larger than the theoretical value of 130 GPa for the (111) planes in [112] orientation (see chapter 5). Factors that further affect the modulus could be due to the inherent porosity, even though it was not observed in these samples (see chapter 5), on the DBs, which are artifacts from the sputter process. A factor that leads to an overestimation of the modulus might be the strain measurement with DIC. If markers loose tracking during analysis and the sample keeps on deforming, but the markers stay at its position, the strain will be underestimated. Consequently the modulus will be overestimated.

In the P-d-method (also referred to as energetic approach) the area underneath the load vs. displacement curve is calculated to determine the applied energy dW. This includes the elastic energy dE_{el} and plastic component dE. The plastic component, which corresponds to the energy needed for crack growth, can be determined by subtracting dE_{el} from dW.

Table 6-3: Various values of fracture toughness determined by the P-d-method (I) and the da/dN vs. ΔK-method (II) (values from [54]) for low density nt Cu (LDNT) and high density nt Cu (HDNT).

specimen	λ (nm)	K_{IC} (MPa$\sqrt{m}$) (P-d)	K_{IC} (MPa$\sqrt{m}$) (da/dN vs. ΔK)
ufg Cu	-	12.9	14.9
LDNT Cu	85 ± 15 nm	14.8	17.5
HDNT Cu	32 ± 7 nm	17.2	22.3

The fracture toughness of a 100 nm thick sample was determined to be on the order of $K_Q = 6.7 \pm 0.8$ MPa$\sqrt{m}$ using the Young's modulus of 168 GPa. As stable crack growth occurred only in the outer twin, a Young's modulus of 130 GPa in the hard (111) orientation in the [112] direction seems to be more realistic. Therefore the calculation seems to be overestimated. After linear elastic fracture mechanics, the conditions for a pure plane stress state are not fulfilled as the sample is too thin [74] and the surface influence leads to a plane strain state which increases the value of the fracture toughness [74]. The plastic zone r_p around the crack tip can be estimated with Eq. (3-2) [74] to be ~ 2.5 µm. The used values for this estimation are the toughness K_Q ~ $7\ MPa\sqrt{m}$ and the yield strength σ_y ~ 1500 MPa. This is a few orders of magnitude greater than the sample thickness. Under these conditions of large scale plasticity the calculated value is not valid fracture toughness K_{IC}.

Nevertheless, the determined value is lower by a factor of two compared to the values determined on nt Cu microsamples with a similar twin density [54] (see Table 6-3). The fracture strain of ε = 2.7 % was also very similar compared to the ε = 2.5 % measured in this work. The yield strength reported by Singh et al. [54] as well as Anderoglu et al. [45] and Hodge et al. [162] and also in chapter *ex situ*, was lower by a factor of 2-3 compared to the results presented in this paper. This extreme increase in strength might partially be explained by the thickness of these samples. The reduction in thickness to 50-100 nm could lead to a strength of σ = 1-2 GPa [193]. It is also suspected that the very small sample volume of the *in situ* TEM samples will lead to an increase in strength. A further going discussion can be found in chapter 6 of the "overall discussion".

6.5.3   Dislocation motion and detwinning

The presumably mobile steps at the CTBs observed in a post test TEM analysis shown in Figure 6-11 are rather similar to what is known from literature as paired partials [56]. Here paired partials (always a multiple of 3) move along the CTBs and therefore shift the position parallel to its normal. This process could be a preliminary stage of detwinning.

The paired partials were only observed in sample ntCu127nmT5 which was deformed at the highest strain rate provided by the setup. The other samples were deformed at a lower strain rate. One reason might be that the fracture or shear edge was not as clearly visible in the TEM after the test and therefore such small features could not be picked up.

The direction of the TBM depends on the local stress state and the twin modification. Under cyclic loading and if the load amplitude is not high enough to push the partial through the whole twin from one domain wall to the next during one loading cycle, the detwinning process is not completed. In this case the partial ledge can move back during unloading and the twin becomes thicker again [52]. The ledge in Figure 6-11 a, seems to be such a multiple of three paired partials moving along the CTB. This corresponds to an ITB segment with a height of roughly $1\,nm$ (in Cu $\left|\vec{b}\right| \sim 0.3\,nm$). The time needed for the detwinning process by partial dislocation motion and local TBM [161] can be estimated by subsequent thinning of the twin. At each cycle a stack of paired partials would move from one domain wall to the next one and reduce the twin thickness by $1\,nm$. In 10 cycles a 10 nm twin could be detwinned. The motion of such ledges works similar to the mechanism of stress coupled GB motion described by Cahn et al. [30]. Kite structures, cells of atoms containing the partials can switch places by a small displacement shape and this process propagates along the CTB like a wave [194, 181]. The speed of this front is expected to reach up to 0.5 -1.2 m/s ($\sim$ 1 µs for the length of 500 nm of the distance between two domain walls) [195], while the mobility is thermally activated and even faster at higher temperatures [181]. Therefore, the detwinning process is not expected to be impeded by a too slow TB mobility.

It is energetically favorable to reduce energy by elimination of CTB area by detwinning [56]. The detwinning process is completed easier for thin twins, as less detwinning partials or a shorter ITB segment have to be involved. As the twins get thinner it gets more difficult to nucleate partials in those twins, but detwinning can also be accomplished from the side of a thicker neighboring twin. Here the probability depends on the constellation of the neighboring twins. This seems to be the situation in Figure 6-11 a, where a CTB of a ~ 5 nm thick twin is containing a ledge, and possibly partial dislocations are contained in the "thicker neighbor". The CTBs that contained detwinning partials appeared to be getting thinner in a stepwise manner, but also seem to be coned (see Figure 6-12).

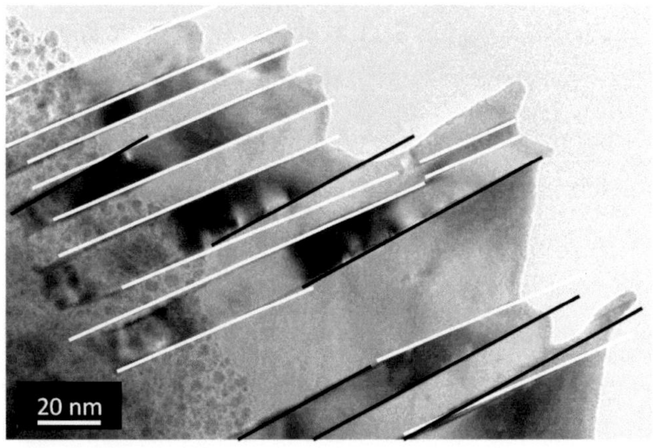

Figure 6-12: BF-TEM micrograph of the fractured/sheared sample ntCu127nmT5 with the CTB traced in blue and red color to indicate the unity of the tilt of the two sets of CTB.

Since the CTBs are all parallel or tilted uniformly this seems to indicate, that either (a) the tilt of the CTB is independent of the number of partial ledges within that CTB, or (b) all CTB contain the same or at least similar number of partial ledges.

In thicker twins like in Figure 6-11 b the diagonally oriented contrast lines are an indication for dislocation activity on $(11\bar{1})$ planes. This might be helpful for accommodation of the applied deformation. Dislocations activity solely on the (111) planes might not be enough.

6.5.4 Nanosample fracture morphology

The fracture morphology of *in situ* nanosamples shows characteristic features and geometries which can be related to the nt structure and their orientation relative to the sample geometry and loading axis. The fracture seemed not to be influenced by the ITB, as fracture was not observed to occur preferentially at the ITB. This is surprising, as the ITB is a preferred location for the formation of pores (Figure 5-13), which can be attributed to the excess volume created during the sputter process. Therefore, the ITBs could be expected to be a site of reduced strength and high fracture probability. This suggests that other mechanisms are more prone to fail the nanosamples.

The samples tested at a low loading rate did not show a strong association to crystallographic orientations in the created fracture edges. In contrast, the thicker ntCu127nmT5 was deformed at an undefined but significantly higher loading rate. The fracture edge seems to be a sheared off region with a consistent width of ~ 60 nm at the fracture surface which was roughly half the sample thickness. The zig-zag fracture along the twin orientation can be related to the crystallographic orientation of the twins which has been seen before in a similar way by Shan et al. [55] and more recently by Kim et al. [156]. However, the zig-zag morphology for thin Cu films was observed before and described as a result of a subsequent shear thinning process by Kim et al. [156]. In that case, the sample thickness might have an effect on the fracture morphology.

The deformation behavior under consideration of dislocation interaction with TBs was recently investigated in molecular dynamic (MD) simulations by T. Zhu et al. at the Georgia Institute of Technology (private communication). The loading axis in the MD simulations was oriented along the [112] direction on the (111) planes. Therefore, the loading situation was identical to the *in situ* experiments conducted in this work. Furthermore, the experimental sample volume was on a similar dimensional magnitude in the simulations and experiments. Due to the enormous computing efforts in MD simulations, only small volumes can be considered. The samples from the experiments are still larger than the volumes of the MD simulations but experimentally as small as seemed reasonable. At least for the considerations of dominant mechanisms, the difference in sample volumes might be an issue. The simulations suggested a shear thinning deformation, triggered by dislocation motion on specific atomic planes, namely the {111} planes (see Figure 6-13). This goes along with a necking in the [110] out-of-plane direction (see Figure 6-14) and is in good agreement with the results observed in our experiments. Due to the different twin modifications, the fracture follows a zig-zag line as was shown clearly in Figure 6-15 a - c and Figure 6-9 b. Interestingly, the very good correlation seems to be only valid for the high strain rate test. Here, it can be argued, that other deformation mechanism than dislocation based deformation are inhibited due to the high strain rate. Therefore, the only mechanism available, is a dislocation based shear thinning process on the {111} plane leading to the aforementioned zig-zag fracture morphology. The angle between the CTB and the fracture surface of ~54° is the same as was determined (see Figure 6-14) in the high strain rate *in situ* experiment and further confirms similitude to the model description. In contrast, the lower strain rate experiments allow for slower processes being involved in the final failure of the samples. E.g. diffusion processes can lead to less pronounced and but rather round shapes at the fracture edges. This explains their rather soft looking fracture morphology at low strain rates.

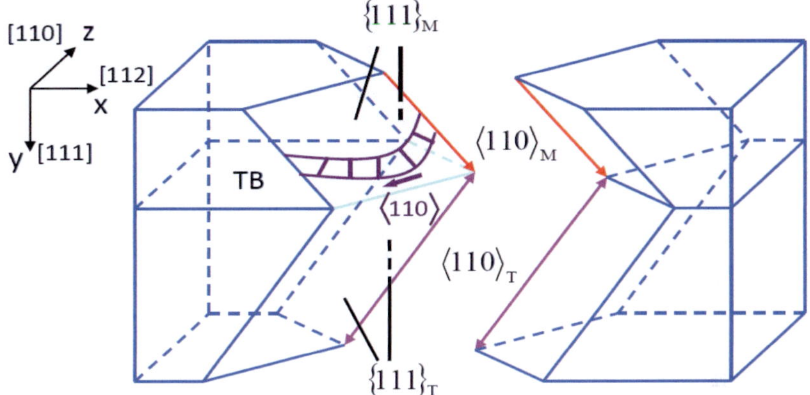

Figure 6-13: Schematic drawing illustrating the fracture morphology by shear thinning. This behavior was motivated by MD simulations (private communication Ting Zhu).

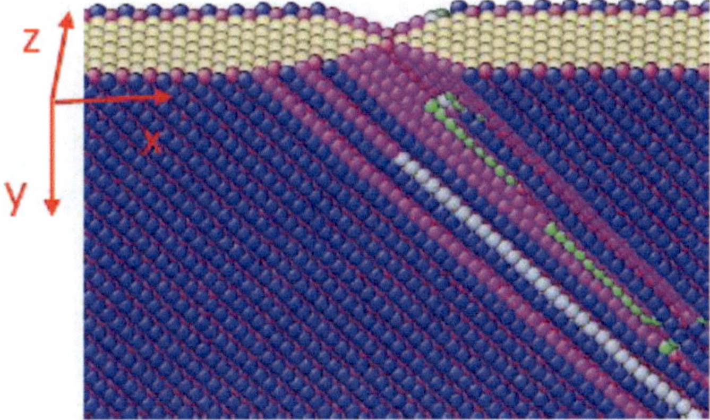

Figure 6-14: Snapshot from MD simulation of the loading of nt material in x direction. The sample is forming a neck at a specific angle depending on the twin modification (private communication Ting Zhu).

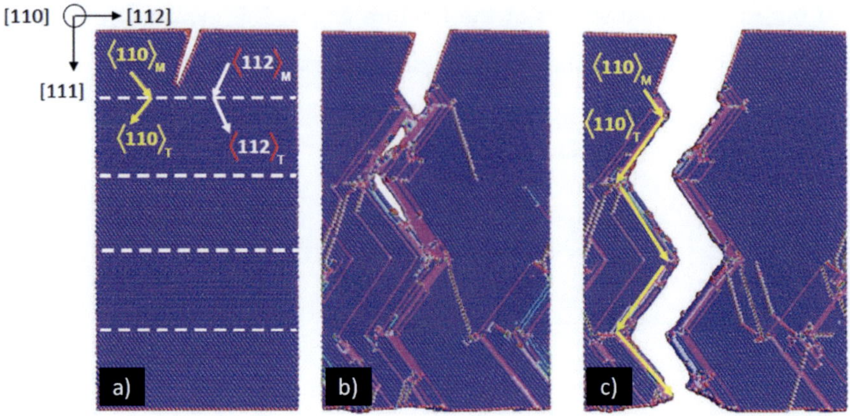

Figure 6-15: (a) Snapshots from MD simulations of a notched nt sample which is loaded horizontally. (b and c) the fractured sample is indicating a fracture angle of ~ 54°. The crack is deflected at the CTB, thereby the zig-zag morphology is correlating to the different twin modification indicated by the index "M" and "T". (private communication Ting Zhu).

6.6   Summary

A new methodology for extracting tensile testing samples of any material and in any orientation has been presented and *in situ* TEM tensile tests were conducted. The tests on those samples of a volume of ~ 10^{-11} mm^3 showed the following results:

1) The samples showed similar high strength of 1.5 – 2.5 GPa and a ductility of ~ 2.5 % in nt and single crystalline Cu. A Young's modulus of 168 GPa ± 10 % and a fracture toughness of 6.7 ± 0.8 MPa√m was measured.

2) While the observed fracture morphology is more irregular at low strain rates, the crystalline character was clearer in the high strain rate test, where the fracture surface is oriented along the twin orientations. At the fracture surface, ledges in form of paired partial dislocations moved along the Σ3 (111) CTB and have the identity of an ITB. Multiples of such

ledges lead to a subsequent twin thinning and finally to detwinning by the motion of a ITB segment. Nevertheless, detwinning was not observed in as the samples were not loaded with a sufficient number of cycles.

3) Fracture never occurred along the ITB but always nearby. At high strain rate, fracture occurred in a zig-zag manner. MD simulations suggested that dislocations that trigger fracture are trapped on {111} planes while diffusional processes were suppressed. Shear thinning, depending on the twin modification led to the observed zig-zag fracture. At low strain rates, fracture occurred with less crystallographic order.

6.7   Acknowledgement

Dipl.-Ing. Jörg Thomas and Dr. Zaoli Zhang is thanked for his help at the TEM in Leoben and Prof. Gerhard Dehm as well as Prof. Ting Zhu for helpful discussion.

7 Summary and concluding remarks

This chapter is based on the results and discussions of the previous chapters. The first two sections are focusing on aspects of microstructural stability and evolution during *in situ* and *ex situ* mechanical loading of nc and nt nano- and microsamples. In the second part the measured yield strength will be discussed in the light of size- and scaling effects. Finally, the possible influence of the machining process on the microstructure and mechanical behavior are discussed.

7.1 Microstructural evolution of nanocrystalline materials

Nanostructured materials are known to have superior mechanical properties compared to their micro structured counterparts. They offer a higher yield and flow stress while retaining a well enough plastic behavior. Nevertheless, the superior properties come at a price, namely the microstructural stability. Three aspects have a strong influence on their behavior as a higher defect density leads to higher driving forces, shorter diffusion paths allow for faster kinetics and the higher strength enables much higher stresses. Therefore, pure nc materials do experience thermally activated grain growth even at RT which diminishes their advanced properties and renders them unpredictable in technical applications. Moreover, the microstructural features are also sensitive to mechanical deformation. Especially repetitive or cyclic loading strongly affects the microstructure and leads to coarsening in nc and nt metals as was observed in literature and shown in the previous chapters. To be able to control this coarsening, a better understanding of the responsible mechanisms is of great interest. Therefore, the following section will focus on the observations and the conclusions which could be drawn to narrow down the active mechanisms leading to coarsening or detwinning.

The nc Ni microstructure coarsened during *ex situ* fatigue experiments in the vicinity of the crack tip and within the plastic zone with a size of $r_p \sim 10\,\mu m$. The degree of coarsening is in agreement with observations by Witney et al. [91] and Padilla et al. [85]. Due to the small grain sizes and short diffusion paths, diffusional processes are expected to be one of the relevant mechanisms of coarsening [196]. This includes dislocation climb, GB sliding and grain rotation. Coarsening did not appear to be continuous, instead a bimodal grain size distribution developed. A similar evolution was observed by Gianola et al. for nc Al [105]. The bimodal structure implies growth of rather large grains. These large grains account for a high volume fraction and therefore have a strong influence on mechanical properties. E.g. in a nc material with an average grain size of $\sim$ 20 nm and 10 % of the grains between 30 and 40 nm, these will represent roughly 50 % of the total volume [40].

During *in situ* cyclic loading in the TEM, a similar evolution of the microstructure was observed in *ex situ* fatigue experiments. A bimodal structure developed in the region of the gage section ($\sim 1\,\mu m$) at various locations. Nevertheless, rotation was not observed *in situ*. Instead, sections of GB moved at locations of high stress, e.g. at the crack tip. This stress coupled GB migration during loading seems to be governing the coarsening behavior and was similarly observed in experiments [105, 123, 197] as well as in molecular dynamic simulations [198] and described theoretically by Cahn et al. [199, 200]. Cahn describes it as an alternative deformation mechanism as the advancement of a GB is coupled to a plastic shear deformation [201]. This process does not require any diffusional mechanism and therefore can lead to a rapid motion of the GBs. This process which drives the coupled GB motion is somewhat similar to a dislocation motion along a GB. The equivalent of the dislocation core would be the position where the GB kite structures are switching between shapes. Nevertheless, it seems intuitive that diffusion based mechanisms are also involved where the material and the grain boundaries are not perfect. The equivalent could be the dislocation climb processes. The observations made here of the fast GB motion indicate that stress coupled GB motion [30] is the main coarsening mechanism in nc Cu and nc Ni.

The *in situ* tested samples showed that failure appeared to be inter- or transcrystalline fracture depending on the grain size. While grains smaller than ~ 100 nm tend to fail intercrystalline and larger grains transcrystalline. An interesting but uncommon observation was made within the nc Cu sample, where in the notch a hard nt Cu grain was located behind a rather soft and plastically deforming nc Cu grain. Such nanotwinned grains were not regular in other samples. While the soft grain sitting at the notch tip just deformed, the stress intensity was transferred to the strong nt Cu grain sitting behind it. The deformation of the soft grain leads to crack bridging, and it dissipated energy while it deformed. At a certain point in the experiment, a crack nucleated and penetrated the hard grain and lead to a stepwise transcrystalline fracture. The crack grew within the nt Cu grain perpendicular to the TB. In this test, the strain in the bridging soft grain was measured to be on the order of ~ 20 %. The situation here is different than in the loading of the whole sample. No other grains surrounded the bridging grain except at the contact points to the crack flanks. Therefore it is not expected to play a role. Furthermore, the testing time for this loading segment was on the order of one second. Diffusion is expected to need more time for the necessary mass transport to provide the large deformation in the soft grain. More likely is that the plastic deformation was induced by dislocation activity but could not be resolved from the TEM images taken during the test

The measured overall strain to failure in this sample was one order of magnitude lower, e.g. $\varepsilon = 2\%$. The high local strength and ductility can be attributed to the fact that in the larger sample volume more defects are available, which can fail and lead to a lower "overall"-strain.

7.2 Microstructural evolution of nanotwinned materials

Similar to the coarsening in nc materials, nt Cu is also undergoing coarsening of the microstructure which is providing its advanced mechanical properties [164, 56, 182]. The coarsening comes along with mechanical softening, which was simulated by Li et al. [57]. The deformation behavior is characterized by an interaction of localized deformation and local detwinning. The whole process is limited to a 20 μm wide shear band with a angle of just under 50° relative to the loading axis. In the center of that shear band detwinning completely removed all nano twins and resulted in an ufg microstructure. Regions with partially detwinned structure connect the transition zone to the initially nt material. In the shear band plastic shear strains of up to 370 % were measured. The elevated strength and "deformability" of nt material would suggest an enhanced fatigue lifetime compared to pure ufg Cu. Due to the softening behavior, the performance during cyclic loading is not significantly better.

TBs are a relatively low energy configuration [202] compared to GBs. Consequently, their thermal stability is higher compared to nc metals [53] and they are not prone to be affected by diffusion and diffusion is not expected to be involved in the detwinning process. However, the detwinning as it was analyzed in the *ex situ* tested samples is triggered by partial dislocation motion on

(111) CTBs. Once the twin is thin enough it can be removed by the motion of an ITB segment in the one or the other direction, depending on the twin modification. The observed interlacing brick-like structure in fatigued nt Cu microsamples is not necessarily correlated to detwinning. The movement of an ITB segment of an unthinned twin from on DB to the other in order to remove the twin would cost too much energy [56]. Nevertheless these interlacing structures could be artifacts of fatigue, but their distance of propagation is limited.

No global detwinning was observed *in situ* as a sufficiently high number of cycles of partial motion along the CTB would be needed. The repetitive loading in the TEM might not have had enough cycles and time for a continuous global detwinning process. Nevertheless, ledges along CTBs and local detwinning at the fracture surface were observed. Due to the volumetric constraints of the nanosamples (size effect) this process took place at higher stresses then in the microsamples and only limited partial dislocation motion as source for detwinning was observed.

7.3 Scaling- and size effects

In Figure 7-1, the yield strengths of the tested samples are presented in comparison to literature values of thin film samples with different film thickness. Except for the nt samples, all films from literature were prepared by physical vapor deposition with a grain size of roughly the film thickness. The Cu thin films from Yu et al. [142] were heavily twinned.

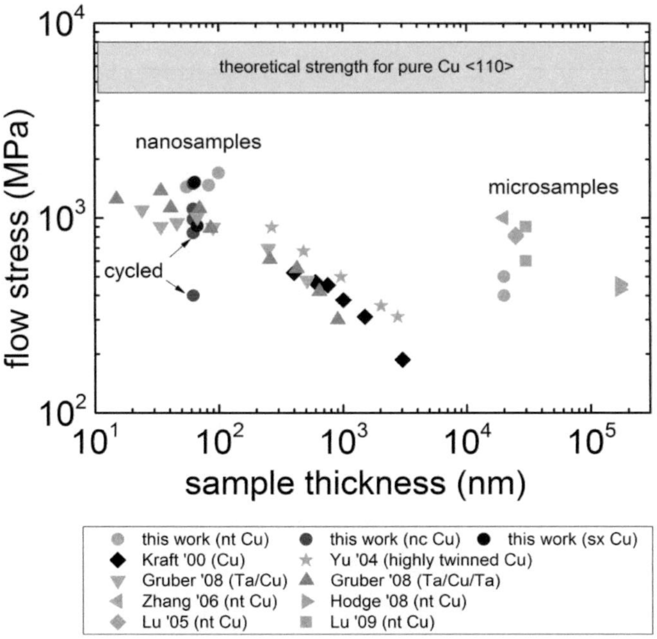

Figure 7-1: Flow stress vs. sample thickness of the tested nt, nc and single crystalline (sx) Cu samples in comparison to literature values from Yu '04 [142], Kraft '00 [143], Gruber '08 [139] Zhang '06 [160] and Hodge '08 [162], Lu '05 [158] and Lu '09 [63].

The yield strength of nt microsamples follows a classical size effect. With a decreasing twin spacing between 100 and 10 nm, the yield strength is increasing between ~ 1000 and 500 MPa respectively [63, 184]. To explain the yield strength as a function of grain size and twin spacing, Gu et al. formulated the non uniform partial dislocation extension model for nc materials [22] and extended it to a generalized model for nc and nt materials [23]. In this model, the flow stress is depending on the SFE, the grain size and twin spacing. With the extension of the SF, a higher stress has to be applied, which correlates to a higher yield strength. The leading and trailing partials that frame the SF can be emitted from existing GB dislocations or stress concentrators [40]. The yield strength of the microsamples is significantly lower than literature values from samples with comparable twin spacing. An explanation for the low strength is the rough and damaged surface caused by EDM cutting. The damage zone in the EDM'ed samples was considerably thicker than in the comparable laser cut samples. This heat affected zone might have triggered early deformation.

The thin film literature values follow the classical scaling effect. Samples with decreasing thickness (grain size ~ film thickness) gain increased yield strength. At ~ 100 nm this effect seems to be exhausted at a strength of ~ 1000 MPa. The explanation for the resulting plateau are detrimental defects that can get activated at this stress level, leading to fracture of these films. This also explains the moderate yield strength compared to the "defect free" whiskers [203].

In the here prepared *in situ* TEM samples, both effects come together. With a film thickness of less than 100 nm the mechanical behavior (strength) is dominated and influenced by the sample volume and film thickness. This implies a constraint of dislocation activity. Accordingly, the inner structure of GBs or TBs has a minor effect on the material strength. Nevertheless, the nt samples have a yield strength slightly higher than the single crystalline and as deposited nc samples, what is an indication for a more effective interaction of scaling- and size effect in the nt material. Additional size constraints that lead to higher stresses enable detwinning partial dislocation activity down to a lower twin thickness. It could be argued, that similar stresses are locally available in larger samples, e.g. in the vicinity of defects or hard grains surrounded by soft grains. Therefore, the *in situ* TEM observations presented here give an insight into what mechanisms are active at critical positions in the thin film.

7.3.1 Influence of the sample processing on microstructure and mechanical properties of *in situ* TEM tensile samples

For preparation and processing of the nanosamples a focused ion beam is used. Using Ga$^+$-ions for machining always induces damage and artifacts. This implies a change in mechanical behavior e.g. embrittlement [130]. Kiener et al. showed that this has a severe effect especially in Cu [129]. One result of this process is GB decohesion [130, 131] resulting in a reduced strength. Other possible consequences are grain growth, depending on which angle the ion-beam is scanned over the surface [111]. This effect could be neglected, as the beam is never focused on the surface under a steep angle. The *in situ* TEM tensile samples showed a roughly ~ 10 nm thick layer of amorphous and redeposited material. The Pt-gas from the gas injection system and contamination with oxides are spread over the whole sample. With increasing thickness, this tends to have a negative influence on the imaging properties in the TEM.

8 Outlook

To increase the fatigue lifetime for nt materials, crack initiation might be impeded by suppression of the detwinning mechanism. Detwinning might be influenced by using alloying atoms to pin the dislocations in the ITB and to prevent their emission on CTBs. This could stabilize the nt microstructure and therefore prevent localization leading to shear band formation and therefore could significantly increase the mechanical performance of the nt material.

To complete the twin thinning process, detwinning partials have to move along the CTB and reach the opposite ITB. Large domains increase the length the partial has to move. A longer distance the partial has to move implies more effort which has to be made to complete the twin thinning and detwinning process.

In terms of small scale testing, improvements could be made on the *in situ* TEM holders since they are only single-tilt holders. Therefore one can tilt the sample only in one direction. To tilt the sample into a zone axis for HRTEM two tilt angles are necessary. To solve that issue, the sample would have to be removed from the PTP chip and mount on a conventional sample holder with two tilt orientations. This is hardly possible with the current PTP geometry. The reason is that in order to unmount the sample holder, the *in situ* indenter tip has to be retracted from the PTP. Since the sample has been deformed and elongated during tensile testing the two separated sample sides get in contact due to the spring load of the PTP. Therefore the interesting region will be destroyed or at least deformed. A system to lock the PTP in its retracted position could help to solve that issue.

Alternatively, the test could be done in an in-situ SEM/FIB system. After the test, one half of the sample would be milled away with the FIB, and the other half could be removed from the PTP and transferred onto a conventional TEM grid. This could then be analyzed in a conventional double tilt TEM holder.

9 Deutsche Zusammenfassung

9.1 Motivation und Literaturüberblick

Die innere Struktur von nanokristallinen (nc) und nanoverzwillingten (nt) Metallen mit einer kleinsten Strukturgröße von weniger als 100 nm führt nach dem Hall-Petch Verhalten durch Behinderung der Versetzungsbewegung zu hervorragenden mechanischen Eigenschaften. Dazu zählen erhöhte Festigkeit und Härte. Diese Eigenschaften sind an die Stabilität der Struktur gebunden und gehen verloren, sobald Vergröberung durch mechanische Belastung eintritt.

Die zugrundeliegenden Mechanismen sind nicht eindeutig geklärt. Dennoch wird vermutet, dass Korngrenzen in nc Meltallen einen wichtigen Anteil an den dominierenden Mechanismen tragen. Nicht nur durch Behinderung der Versetzungsbewegung und der alternativ ablaufenden Partialversetzungs-bewegung, sondern auch als direkt involviert gelten Korngrenzen bei Kornrotation und KG-Abgleiten (Koaleszenz). Desweiteren spielt aufgrund der kleinen Korngröße Diffusion entlang von Korngrenzen eine wichtige Rolle. Ein weiterer Mechanismus welcher in der Literatur zur Verformung nc Metalle Erwähnung findet ist spannungsinduzierte Korngrenzmigration. Hierbei bewegen sich Korngrenzen durch spannungsinduzierte Umklappvorgänge entlang der Korngrenze.

Für das globale Verhalten der Proben wurden *ex situ* Versuche and Mikroproben mit einer Dicke von bis zu 200 µm durchgeführt. *In situ* Versuche im Transmissionselektronenmikroskop (TEM) an Nanoproben mit einer Dicke von bis zu ~ 100 nm dienten dazu, lokale Verformungsmechanismen beobachten zu können. Als Material für die Versuche an Mikroproben wurde nc Ni und nt Cu verwendet, für die Versuche an Nanoproben nc und nt Cu.

9.2 Experimentelles

Die kommerziell erworbenen 200 µm dicken nc Ni Folien wurden elektrochemisch abgeschieden und dann mit Drahterodieren in die gewünschte Probenform geschnitten.

Nc Cu wurde für die *in situ* Versuche in einer Dicke von ~ 50 nm auf Plättchen aus Schichtssilikat magnetron gesputtert. Durch das Einlegen in Wasser löste sich die Schicht vom Substrat ab und schwamm auf der Wasseroberfläche. Für das Testen kamen sog. „Push-to-pull devices" (PTP) zum Einsatz. Mit diesen konnte man den Film aus nc Cu von der Wasseroberfläche abheben. Die Aufgabe des PTP bestand nun darin, die Druckkraft aus dem Indenter des *in situ* TEM Probenhalters in eine Zugkraft auf der Probe umzuwandeln.

Das nt Cu wurde ebenfalls magnetron gesputtert und von einer Gruppe der University of Texas zur Verfügung gestellt. Der ursprüngliche Film hat eine Dicke von 20 µm. Daraus wurden zum Einen durch Drahterodieren die Mikroproben hergestellt. Zum Anderen wurde eine Methode entwickelt um durch fokusierte Ionenstrahlen die nt Cu Nanoproben herzustellen. Dabei werden TEM Lamellen mit etwa 10 µm langen Zugarmen geschnitten, im mittleren zu testenden Bereich auf elektronentransparenz gedünnt und waagrecht auf einen PTP abgelegt. Anschließend erfolgt noch das genaue Zuschneiden der gewünschten Probengeometrie im Testbereich. Mit dieser Methode können Zugproben beliebiger Orientierung präpariert werden.

Die *ex situ* Versuche an den Mikroproben erfolgten in einem eigens entwickeltem und gebauten Aufbau. *In situ* TEM (JEOL 2010, Japan) erfolgte in mithilfe eines speziellen Halters, in welchen ein Indenter zur Erzeugung der mechanischen Last integriert war.. Außerdem erfolgte im abgescherten Bereich einer nt Cu Probe lokale Bestimmung der Orientierungsbeziehungen mit „Orientation mapping" (OIM).

9.3 Ergebnisse und Diskussion

Bei den *ex situ* Ermüdungsversuchen an nc Ni trat im plastischen Einflussbereich von ein paar µm um die Rissspitze eine lokale Vergröberung bzw. Bimodalisierung des nc Gefüges auf. Dabei konnte Versetzungsaktivität in bis zu 30 nm kleinen Körnern festgestellt werden. Für kleinere Körner gelten spannungsinduzierte Korngrenzbewegung und Diffusion als dominante Verformungs - und Vergröberungsmechanismen.

Eine ähnliche Vergröberung trat bei den nc Cu Nanoproben während in situ Belastung im TEM auf. Als Mechanismen werden ebenfalls Diffusion und spannungsinduzierte Korngrenzbewegung vermutet. Korngröße ab 100 nm, bis 140 nm transkristallin und interkristallin. In einem nt Korn konnte stufenweises Bruchverhalten einzelner Zwillingsligamente beobachtet werden. Dabei blieben „weichere" und nicht nt und von der Rißspitze weitere entfernte Komponenten der Mikrostruktur intakt. Dies führte zu sog. „crack bridging" und stellt einen möglichen Mechanismus zur Zähigkeitssteigerung dar. Da die Mikrostruktur einphasig ist, unterscheidet es sich zu konventionellem „bridging" bei dem es sich meistens um eine zweiphasige Mikrostruktur handelt. Die gemessene Festigkeit von bis zu 2.5 GPa ist auf das geringe Volumen und die damit einhergehende geringe Defektdichte zurückzuführen.

Bei den *ex situ* Versuchen an nt Cu Makroproben viel zuerst die lokale Scherverformung von bis zu 370 % entlang eines eta 15 µm breiten

Scherbandes auf. In diesem Band zeigten TEM Untersuchungen eine Auflösung der Zwillinge und Vergröberung der Mikrostruktur. Desweiteren stellte eine Analyse mit „orienataion mapping" im TEM den Übergang von ursprünglichem und unverändertem nt Gefüge hin zum ultra feinkörnigem (ufg) Gefüge im Scherbandzentrum dar. Dies erfolgt nicht nur durch Vergröberung sondern zusätzliche Rotationsprozesse verleihen dem ufg Gefüge polykristallinen Charakter. Die Vergröberung der Zwillingsstruktur erfolgt im ersten Schritt durch Bewegung von gepaarten Partialversetzungen entlang der kohärenten $\Sigma 3$ Zwillingsgrenzen auf (111) Ebenen. Durch diesen Prozess sinkt die Zwillingsbreite der einen Zwillingsmodifikation. Sind die Zwillinge dünn genug ist es energetisch günstiger einen kompletten Zwilling durch die Bewegung einer inkohärenten Zwillingsgrenze in die Modifikation der benachbarten Zwillinge umzuwandeln. Bei ausreichender Strukturgröße setzen konventionelle Versetzungsbewegung ein und erlaubt Rotation von Gefügestrukturen. Desweiteren fiel als mechanischer Aspekt bei den Ermüdungsversuchen der Frequenzeinfluss auf. Somit zeigen Proben bei höherer Ermüdungsfrequenz eine längere Lebensdauer. Dies steht im Einklang mit Literaturwerten von Versuchen an ursprünglichen ufg Cu Proben, welche im Ermüdungsschaubild mit den Werten von nt Cu zur Deckung kommen. Eine mögliche Erklärung hierfür ist, dass zwar durch die erhöhte mechanische Festigkeit und moderate Duktilität von nt Cu zwar eine erhöhte Ermüdungsfestigkeit zu erwarten wäre, dennoch ist die Verformung sehr lokalisiert, was die Ermüdungsfestigkeit wiederum verschlechtert.

In situ TEM Versuche an nt Cu Nanoproben zeigten mit maximal 2.6 GPa ähnliche Festigkeiten wie nc Cu. In Abhängigkeit der Dehnrate ergab sich ein unterschiedliches Bruchbild. Im Gegensatz zur Belastung bei geringen Dehnraten ist das Bruchbild bei hoher Dehnrate gezeichnet von hoher kristalliner Ordnung. Durch die kurze Dauer der Belastung sind Diffusion und Klettervorgänge unterdrückt. Versetzungbewegung konzentriert sich auf {111} Ebenen welche zu Dünnung durch Scherverformung führt. Abhängig von der Zwillingsmodifikation bildet sich schließlich ein zick-zack Bruch mit Bruchflanken unter ~54° zur Zugachse. Molekulardynamik Simulationen an nt Cu der Gruppe von Ting Zhu des Georgia Institute of Technology unterstützen dieses Modell.

10 References

[1]   www.xtalic.com. December, 2011.

[2]   www.cut tools.de. November 2011.

[3]   de.wikipedia.org/wiki/Mikrochip. December 2011.

[4]   thumbs.dreamstine.com/thumblarge_496/1271615987klv8lA.jpg. December 2011.

[5]   www.i65chromeshop.com. November 2011.

[6]   R. Feynman. There is plenty room at the bottom. *APS Annual Meeting, Caltech*, 1959.

[7]   H. Gleiter. Nanocrystalline materials. *Progress in Materials Science*, 33(4):223 – 315, 1989.

[8]   K. S. Kumar, H. Van Swygenhoven, and S. Suresh. Mechanical behavior of nanocrystalline metals and alloys. *Acta Materialia*, 51(19):5743 – 5774, 2003. The Golden Jubilee Issue. Selected topics in Materials Science and Engineering: Past, Present and Future.

[9]   M.A. Meyers, A. Mishra, and D.J. Benson. Mechanical properties of nanocrystalline materials. *Progress in Materials Science*, 51(4):427 – 556, 2006.

[10]   C. Koch. Structural nanocrystalline materials: an overview. *Journal of Materials Science*, 42(5):1403–1414, 2007.

[11]   F. Dalla Torre, H. Van Swygenhoven, and M. Victoria. Nanocrystalline electrodeposited Ni: microstructure and tensile properties. *Acta Materialia*, 50(15):3957 – 3970, 2002.

[12]   H. Gleiter. Nanostructured materials: basic concepts and microstructure. *Acta Materialia*, 48(1):1 – 29, 2000.

[13]   N. J. Petch. *J. Iron Steel Inst*, 173:25, 1951.

[14]   E. O. Hall. *Proc. Phys. Soc*, 643:747, 1951.

[15]   N. J. Petch. The cleavage strength of polycrystals. pages 25 −28, 1953.

[16]   J. R. Trelewicz and C. A. Schuh. The Hall-Petch breakdown in nanocrystalline metals: A crossover to glass-like deformation. *Acta Materialia*, 55(17):5948 − 5958, 2007.

[17]   C.E. Carlton and P.J. Ferreira. What is behind the inverse Hall-Petch effect in nanocrystalline materials? *Acta Materialia*, 55(11):3749 − 3756, 2007.

[18]   G.J. Fan, H. Choo, P.K. Liaw, and E.J. Lavernia. A model for the inverse Hall-Petch relation of nanocrystalline materials. *Materials Science and Engineering: A*, 409(1-2):243 − 248, 2005. Micromechanics of Advanced Materials II - TMS 2005 Annual Meeting, in Honour of James C.M. Li's 80th Birthday.

[19]   J.R. Weertman. Hall-petch strengthening in nanocrystalline metals. *Materials Science and Engineering: A*, 166(1-2):161–167, July 1993.

[20]   E. Arzt. Size effects in materials due to microstructural and dimensional constraints: a comparative review. *Acta Materialia*, 46(16):5611 − 5626, 1998.

[21]   T. Volpp, E. Göring, W. M. Kuschke, and E. Arzt. Grain size determination and limits to Hall-Petch behavior in nanocrystalline nial powders. *Nanostructured Materials*, 8(7):855–865, October 1993.

[22]   P. Gu, B. K. Kad, and M. Dao. A modified model for deformation via partial dislocations and stacking faults at the nanoscale. *Scripta Materialia*, 62(6):361–364, March 2010.

[23]   P. Gu, M. Dao, R. J. Asaro, and S. Suresh. A unified mechanistic model for size-dependent deformation in nanocrystalline and nanotwinned metals. *Acta Materialia*, 59(18):6861–6868, October 2011.

[24]   Y. Furuya. Specimen size effects on gigacycle fatigue properties of high-strength steel under ultrasonic fatigue testing. *Scripta Materialia*, 58(11):1014 – 1017, 2008.

[25]   G. G. Trantina. Statistical fatigue failure. *Journal of Testing and Evaluation*, 9:44–49, 1981.

[26]   G. E. Dieter. *Mechanical Metallurgy*. McGraw-Hill, 1986.

[27]   M. D. Uchic, D. M. Dimiduk, J. N. Florando, and W. D. Nix. Sample dimensions influence strength and crystal plasticity. *Science*, 305(5686):986–989, August 2004.

[28]   M. R. Fitzsimmons, J. A. Eastman, M.Müller-Stach, and G. Wallner. Structural characterization of nanometer-sized crystalline Pd by x-ray-diffraction techniques. *Phys. Rev. B*, 44(6):2452–, August 1991.

[29]   E. A. Stern, R. W. Siegel, M. Newville, P. G. Sanders, and D. Haskel. Are nanophase grain boundaries anomalous? *Phys. Rev. Lett.*, 75(21):3874–, November 1995.

[30]   J. W. Cahn, Y. Mishin, and A. Suzuki. Coupling grain boundary motion to shear deformation. *Acta Materialia*, 54(19):4953 – 4975, 2006.

[31]   Y. Kuru, M. Wohlschlögel, U. Welzel, and E.J. Mittemeijer. Large excess volume in grain boundaries of stressed, nanocrystalline metallic thin films: Its effect on grain-growth kinetics. *Applied Physics Letters*, 95, 2009.

[32]   R. K. Guduru, K. L. Murty, K. M. Youssef, R. O. Scattergood, and C. C. Koch. Mechanical behavior of nanocrystalline copper. *Materials Science and Engineering: A*, 463(1-2):14–21, August 2007.

[33]   J. Aktaa, J.Th. Reszat, M. Walter, K. Bade, and K.J. Hemker. High cycle fatigue and fracture behavior of LIGA nickel. *Scripta Materialia*, 52(12):1217 – 1221, 2005.

[34] Y.J. Li, J. Mueller, H.W. Höppel, M. Göken, and W. Blum. Deformation kinetics of nanocrystalline nickel. *Acta Materialia*, 55(17):5708 – 5717, 2007.

[35] C. C. Koch. Optimization of strength and ductility in nanocrystalline and ultrafine grained metals. *Scripta Materialia*, 49(7):657–662, October 2003.

[36] X. Shen, J. Lian, Z. Jiang, and Q. Jiang. High strength and high ductility of electrodeposited nanocrystalline Ni with a broad grain size distribution. *Materials Science and Engineering: A*, 487(1-2):410 – 416, 2008.

[37] X. Shen, J. Lian, Z. Jiang, and Q. Jiang. The optimal grain sized nanocrystalline Ni with high strength and good ductility fabricated by a direct current electrodeposition. *Adv. Eng. Mater.*, 10(6):539–546, 2008.

[38] Y. Zhao, X. Liao, S. Cheng, E. Ma, and Y. Zhu. Simultaneously increasing the ductility and strength of nanostructured alloys. *Adv. Mater.*, 18(17):2280–2283, 2006.

[39] L. Hollang, E. Hieckmann, C. Holste, and W. Skrotzki. Strain-rate sensitivity of additive-free pulsed-electrodeposited nickel during cyclic loading. *Materials Science and Engineering: A*, 483-484:406 – 409, 2008. 14th International Conference on the Strength of Materials.

[40] R. J. Asaro and S. Suresh. Mechanistic models for the activation volume and rate sensitivity in metals with nanocrystalline grains and nano-scale twins. *Acta Materialia*, 53(12):3369–3382, July 2005.

[41] Q Wei, S Cheng, K.T Ramesh, and E Ma. Effect of nanocrystalline and ultrafine grain sizes on the strain rate sensitivity and activation volume: fcc versus bcc metals. *Materials Science and Engineering: A*, 381(1â€"2):71–79, September 2004.

[42]   O. Anderoglu, A. Misra, F. Ronning, H. Wang, and X. Zhang. Significant enhancement of the strength-to-resistivity ratio by nanotwins in epitaxial Cu films. *Journal of Applied Physics*, 106, 2009.

[43]   L. Lu, X. Chen, X. Huang, and K. Lu. Revealing the maximum strength in nanotwinned copper. *Science*, 323(5914):607–610, January 2009.

[44]   S. Dahlgren. Columnar grains and twins in high-purity sputtere-deposited copper. *Journal of Vacuum Science and Technology*, 11:832 – 836, 1974.

[45]   O. Anderoglu, A. Misra, H. Wang, F. Ronning, M. F. Hundley, and X. Zhang. Epitaxial nanotwinned Cu films with high strength and high conductivity. *Applied Physics Letters*, 93(8):083108, August 2008.

[46]   H. Jiang, T.J. Klemmer, J.A. Barnard, W.D. Doyle, and E.A. Payzant. Epitaxial growth of Cu (111) films on Si (110) by magnetron sputtering: orientation and twin growth. *Thin Solid Films*, 315(1-2):13–16, March 1998.

[47]   Y. B. Wang, M. L. Sui, and E. Ma. In situ observation of twin boundary migration in copper with nanoscale twins during tensile deformation. *Philosophical Magazine Letters*, 87(12):935–942, November 2007.

[48]   A. Misra. Twinning in nanocrystalline metals. *JOM Journal of the Minerals, Metals and Materials Society*, 60(9):59–59, 2008-09-01.

[49]   D. Bufford, H. Wang, and X. Zhang. High strength, epitaxial nanotwinned Ag films. *Acta Materialia*, 59(1):93–101, January 2011.

[50]   O. Anderoglu, A. Misra, H. Wang, and X. Zhang. Thermal stability of sputtered Cu films with nanoscale growth twins. *Journal of Applied Physics*, 103(9):094322 1–6, May 2008.

[51]   X. Zhang, O. Anderoglu, R. Hoagland, and A. Misra. Nanoscale growth twins in sputtered metal films. *JOM Journal of the Minerals, Metals and Materials Society*, 60(9):75–78, 2008.

[52]   N. Li, J. Wang, A. Misra, X. Zhang, J.Y. Huang, and J.P. Hirth. Twinning dislocation multiplication at a coherent twin boundary. *Acta Materialia*, 59(15):5989–5996, September 2011.

[53]   O. Anderoglu, A. Misra, H. Wang, and X. Zhang. Thermal stability of sputtered Cu films with nanoscale growth twins. *Journal of Applied Physics*, 103(9):094322, May 2008.

[54]   A. Singh, L. Tang, M. Dao, L. Lu, and S. Suresh. Fracture toughness and fatigue crack growth characteristics of nanotwinned copper. *Acta Materialia*, 59(6):2437–2446, April 2011.

[55]   Z. W. Shan, L. Lu, A. Minor, E. Stach, and S. X. Mao. The effect of twin plane spacing on the deformation of copper containing a high density of growth twins. *JOM*, 60:71 – 74, 2008.

[56]   Y.B. Wang and M.L.Sui. In situ transmission electron microscopy investigation of the deformation behavior of Cu with nanoscale twins. *Materials Science Forum*, pages 63–72, 2010.

[57]   X. Li, Y. Wei, L. Lu, K. Lu, and H. Gao. Dislocation nucleation governed softening and maximum strength in nano-twinned metals. *Nature*, 464(7290):877–880, April 2010.

[58]   M. D. Merz and S. D. Dahlgren. Tensile strength and work hardening of ultrafine grained high purity copper. *Journal of Applied Physics*, 46(8):3235 – 3237, 1975.

[59]   Z.-H. Jin, P. Gumbsch, E. Ma, K. Albe, K. Lu, H. Hahn, and H. Gleiter. The interaction mechanism of screw dislocations with coherent twin boundaries in different face-centred cubic metals. *Scripta Materialia*, 54(6):1163–1168, March 2006.

[60]   Z.-H. Jin, P. Gumbsch, K. Albe, E. Ma, K. Lu, H. Gleiter, and H. Hahn. Interactions between non-screw lattice dislocations and coherent twin boundaries in face-centered cubic metals. *Acta Materialia*, 56(5):1126–1135, March 2008.

[61]   Z.X. Wu, Y.W. Zhang, and D.J. Srolovitz. Dislocation-twin interaction mechanisms for ultrahigh strength and ductility in nanotwinned metals. *Acta Materialia*, 57(15):4508–4518, September 2009.

[62]   M. Dao, L. Lu, Y.F. Shen, and S. Suresh. Strength, strain-rate sensitivity and ductility of copper with nanoscale twins. *Acta Materialia*, 54(20):5421–5432, December 2006.

[63]   K. Lu, L. Lu, and S. Suresh. Strengthening materials by engineering coherent internal boundaries at the nanoscale. *Science*, 324(5925):349–352, April 2009.

[64]   O. Anderoglu, A. Misra, J. Wang, R.G. Hoagland, J.P. Hirth, and X. Zhang. Plastic flow stability of nanotwinned Cu foils. *International Journal of Plasticity*, 26(6):875–886, June 2010.

[65]   Z. Shan and S. Mao. Direct evidence of a deformation mechanism crossover in nanocrystalline nickel. *Adv. Eng. Mater.*, 7(7):603–606, 2005.

[66]   Y.T. Zhu and T.G. Langdon. Influence of grain size on deformation mechanisms: An extension to nanocrystalline materials. *Materials Science and Engineering: A*, 409(1-2):234–242, November 2005.

[67]   Z. Shan, E. A. Stach, J. M. K. Wiezorek, J. A. Knapp, D. M. Follstaedt, and S. X. Mao. Grain boundary-mediated plasticity in nanocrystalline nickel. *Science*, 305(5684):654–657, July 2004.

[68]   K. E. Harris, V. V. Singh, and A. H. King. Grain rotation in thin films of gold. *Acta Materialia*, 46(8):2623 – 2633, 1998.

[69]   E. Ma. watching the nanograins roll. *science*, 305:623–624, 2004.

[70]   K.R. Magid, R.D. Nyilas, and R. Spolenak. Metal plasticity by grain rotation–microdiffraction case studies. *Materials Science and Engineering: A*, 524(1-2):33–39, October 2009.

[71]   V. Yamakov, D. Wolf, S.R. Phillpot, A.K. Mukherjee, and H. Gleiter. Deformation-mechanism map for nanocrystalline metals by molecular-dynamics simulation. *nature*, 3:43 − 47, 2004.

[72]   S. Divinski. Diffusion in nanostructured materials. *Defect and Diffusion Forum*, pages 289 − 292, 2009.

[73]   G. Gottstein. Physikalische Grundlagen der Materialkunde. Springer, 2001.

[74]   J. Rösler, H. Harders, and M. Bäker. Mechanisches Verhalten der Werkstoffe. 2008.

[75]   J.R. Weertman and P. G. Sanders. Plastic deformation of nanocrystalline metals. *Solid State Phenomena*, 35-36:249–262, 1993.

[76]   R. Birringer. Nanocrystalline materials. *Materials Science and Engineering A*, 117:33 − 43, 1989.

[77]   RJ. Asaro, P. Krysal, and B. Kad. Deformation mechansim transitions in nanoscale fcc metals. *Philos Mag Lett*, pages 733–43, 2003.

[78]   Van H. Swygenhoven and P. M. Derlet. Grain-boundary sliding in nanocrystalline fcc metals. *Phys. Rev. B*, 64(22):224105–, November 2001.

[79]   B. Zhang, K.H. Sun, Y. Liu, and G.P. Zhang. On the length scale of cyclic strain localization in fine-grained copper films. *Philosophical Magazine Letters*, 90(1):69–76, December 2009.

[80]   F. Mompiou, D. Caillard, and M. Legros. Grain boundary shear-migration coupling–I. in situ tem straining experiments in al polycrystals. *Acta Materialia*, 57(7):2198–2209, April 2009.

[81]   M. F. Ashby. The deformation of plastically non-homogeneous materials. 1969.

[82]   K. S. Kumar, S. Suresh, M. F. Chisholm, J. A. Horton, and P. Wang. Deformation of electrodeposited nanocrystalline nickel. *Acta Materialia*, 51(2):387 – 405, 2003.

[83]   H. Mughrabi and H. W. Höppel. Cyclic deformation and fatigue properties of very fine-grained metals and alloys. *International Journal of Fatigue*, 32(9):1413–1427, September 2010.

[84]   S. Suresh. *fatigue of materials*. university press, Cambridge 2nd edition, 1999.

[85]   H. Padilla and B. Boyce. A review of fatigue behavior in nanocrystalline metals. *Experimental Mechanics*, 50(1):5–23–23, 2010.

[86]   A. Weidner, R. Beyer, C. Blochwitz, C. Holste, A. Schwab, and W. Tirschler. Slip activity of persistent slip bands in polycrystalline nickel. *Materials Science and Engineering: A*, 435-436:540 – 546, 2006.

[87]   T. Hanlon, Y. N. Kwon, and S. Suresh. Grain size effects on the fatigue response of nanocrystalline metals. *Scripta Materialia*, 49(7):675 – 680, 2003. Viewpoint Set No. 31. Mechanical Properties of Fully Dense Nanocrystalline Metals.

[88]   T. Hanlon, E.D. Tabachnikova, and S. Suresh. Fatigue behavior of nanocrystalline metals and alloys. *International Journal of Fatigue*, 27(10-12):1147 – 1158, 2005. Fatigue Damage of Structural Materials V.

[89]   H. Mughrabi. Dislocation wall and cell structures and long-range internal stresses in deformed metal crystals. *Acta Metallurgica*, 31(9):1367–1379, September 1983.

[90]   S. R. Agnew and J. R. Weertman. Cyclic softening of ultrafine grain copper. *Materials Science and Engineering A*, 244(2):145 – 153, 1998.

[91]   A. B. Witney, P. G. Sanders, J. R. Weertman, and J. A. Eastman. Fatigue of nanocrystalline copper. *Scripta Metallurgica et Materialia*, 33(12):2025 – 2030, 1995.

[92]   B. Moser, T. Hanlon, K.S. Kumar, and S. Suresh. Cyclic strain hardening of nanocrystalline nickel. *Scripta Materialia*, 54(6):1151 – 1155, 2006. Viewpoint set no. 40: Grain boundary engineering.

[93]   T. Connolley, P. E. MchUgh, and M. Bruzzi. A review of deformation and fatigue of metals at small size scales. *Fatigue & Fracture of Engineering Materials & Structures*, 28(12):1119–1152, 2005.

[94]   M. A. Haque and M. T. A. Saif. Application of MEMS force sensors for in situ mechanical characterization of nano-scale thin films in sem and tem. *Sensors and Actuators A: Physical*, 97-98:239 – 245, 2002.

[95]   M. Haque and M. Saif. In-situ tensile testing of nano-scale specimens in SEM and TEM. *Experimental Mechanics*, 42(1):123–128, 2002.

[96]   M.T.A. Saif M.A. Haque. In situ tensile testing of nanoscale freestanding thin films inside a transmission electron microscope. *J. Mater. Res.*, 20:1769–1777, 2005.

[97]   Yong Zhu and Horacio D. Espinosa. An electromechanical material testing system for in situ electron microscopy and applications. *Proceedings of the National Academy of Sciences of the United States of America*, 102(41):14503–14508, October 2005.

[98] H.D. Espinosa, Yong Zhu, and N. Moldovan. Design and operation of a mems-based material testing system for nanomechanical characterization. *Microelectromechanical Systems, Journal of DOI - 10.1109/JMEMS.2007.905739*, 16(5):1219–1231, 2007.

[99] H. Guo, K. Chen, Y. Oh, K. Wang, C. Dejoie, S. A. Syed Asif, O. L. Warren, Z. W. Shan, J. Wu, and A. M. Minor. Mechanics and dynamics of the strain-induced M1-M2 structural phase transition in individual Vo$_2$ nanowires. *Nano Lett.*, 11(8):3207–3213, July 2011.

[100] S. R. Agnew, B. R. Elliott, C. J. Youngdahl, K. J. Hemker, and J. R. Weertman. Microstructure and mechanical behavior of nanocrystalline metals. *Materials Science and Engineering A*, 285(1-2):391 – 396, 2000.

[101] H. Gleiter. Diffusion in nanostructured metals. *phys. stat. sol. (b)*, 172(1):41–51, 1992.

[102] Y.R. Kolobov, G.P. Grabovetskaya, K.V. Ivanov, and M.B. Ivanov. Grain boundary diffusion and mechanisms of creep of nanostructured metals. *Interface Science*, 10(1):31–36, 2002.

[103] H. Mughrabi and H. Hoeppel. Cyclic deformation and fatigue properties of ultrafine grain size materials: current status and some criteria for improvement of the fatigue resistance. *Mater Res Soc Symp Proc*, 634:B2.1.1.–B2.1.12., 2001.

[104] H. Mughrabi. *On the grain-size dependence of metal fatigue: outlook on the fatigue of ultrafine-grained materials, Investigations and applications of severe plastic deformation*. NATO Science Series, 2000.

[105] D.S. Gianola, S. Van Petegem, M. Legros, S. Brandstetter, H. Van Swygenhoven, and K.J. Hemker. Stress-assisted discontinuous grain

growth and its effect on the deformation behavior of nanocrystalline aluminum thin films. *Acta Materialia*, 54(8):2253–2263, May 2006.

[106] S. Meister, J. Lohmiller, and M Funk. line intersection function. pages http://www.mathworks.com/matlabcentral/fileexchange/35203–line–cut, February 2012.

[107] J.-P. Royet. Stereology: a method for analyzing images. *Progress in Neurobiology*, 37:433 – 474, 1991.

[108] D. Son, J.-J. Kim, T. Won Lim, and D. Kwon. Evaluation of fatigue strength of LIGA nickel film by microtensile tests. *Scripta Materialia*, 50(10):1265 – 1269, 2004.

[109] P. Cavaliere. Fatigue properties and crack behavior of ultra-fine and nanocrystalline pure metals. *International Journal of Fatigue*, 31(10):1476 – 1489, 2009.

[110] G.P. Zhang, C.A. Volkert, R. Schwaiger, P. Wellner, E. Arzt, and O. Kraft. Length-scale-controlled fatigue mechanisms in thin copper films. *Acta Materialia*, 54(11):3127 – 3139, 2006.

[111] R. Spolenak, L. Sauter, and C. Eberl. Reversible orientation-biased grain growth in thin metal films induced by a focused ion beam. *Scripta Materialia*, 53(11):1291–1296, December 2005.

[112] Yu.R. Kolobov, G.P. Grabovetskaya, K.V. Ivanov, A.P. Zhilyaev, and R.Z. Valiev. Grain boundary diffusion characteristics of nanostructured nickel. *Scripta Materialia*, pages 873–878, 2001.

[113] J. Horváth, R. Birringer, and H. Gleiter. Diffusion in nanocrystalline material. *Solid State Communications*, 62(5):319–322, May 1987.

[114] S. P. Joshi and K.T. Ramesh. Rotational diffusion and grain size dependent shear instability in nanostructured materials. *Acta Materialia*, 56(2):282–291, January 2008.

[115] S.V. Bobylev, A.K. Mukherjee, and Ovid'ko I.A. Transition from plastic shear into rotation deformation mode in nanocrystalline metals and cermics. *Rev.Adv.Mater.Sci.*, 19:103–113, 2009.

[116] Y. Yang, B.I. Imasogie, S.M. Allameh, B. Boyce, K. Lian, J. Lou, and W.O. Soboyejo. Mechanisms of fatigue in LIGA Ni MEMS thin films. *Materials Science and Engineering: A*, 444(1-2):39 – 50, 2007.

[117] T. Hanlon, E.D. Tabachnikova, and S. Suresh. Fatigue behavior of nanocrystalline metals and alloys. *International Journal of Fatigue*, 27:1147–1158, 2005.

[118] L. Lu, Y. Shen, X. Chen, L. Qian, and K. Lu. Ultrahigh strength and high electrical conductivity in copper. *Science*, 304(5669):422–426, April 2004.

[119] J. Weissmüller and J. Markmann. Deforming nanocrystalline metals: New insights, new puzzles. *Advanced Engineering Materials Science and Engineering A*, 7:202 – 207, 2005.

[120] V. Yamakov, D. Wolf, S. R. Phillpot, and H. Gleiter. Grain-boundary diffusion creep in nanocrystalline palladium by molecular-dynamics simulation. *Acta Materialia*, 50(1):61–73, January 2002.

[121] D. Kiener, C. Motz, and G. Dehm. Dislocation-induced crystal rotations in micro-compressed single crystal copper columns. *J. Mater. Sci*, 43:2503–2506, 2008.

[122] H. Van Swygenhoven and P. M. Derlet. Grain-boundary sliding in nanocrystalline fcc metals. *Phys. Rev. B*, 64(22):224105–, November 2001.

[123] M. Legros, D. S. Gianola, and K. J. Hemker. In situ tem observations of fast grain-boundary motion in stressed nanocrystalline aluminum films. *Acta Materialia*, 56(14):3380–3393, August 2008.

[124] K.A. Padmanabhan, G.P. Dinda, H. Hahn, and H. Gleiter. Inverse Hall-Petch effect and grain boundary sliding controlled flow in nanocrystalline materials. *Materials Science and Engineering: A*, 452-453:462–468, April 2007.

[125] A. Froseth, P. Derlet, and H. Van Swygenhoven. Twinning in nanocrystalline fcc metals. *Adv. Eng. Mater.*, 7(1-2):16–20, 2005.

[126] D.V. Bachurin and P. Gumbsch. Accommodation processes during deformation of nanocrystalline palladium. *Acta Materialia*, 58(16):5491–5501, September 2010.

[127] W. C. Overton Jr. and J. Gaffney. Temperature variation of the elastic constants of cubic elements. I. copper. *Phys. Rev.*, 98:969–77, 1955.

[128] G. Simmons and H. Wang. *Single Crystal Elastic Constants and Calculated Aggregate Properties: A Handbook*. Cambridge, MA: MIT Press, 1971.

[129] D. Kiener, C. Motz, M. Rester, M. Jenko, and G. Dehm. Fib damage of Cu and possible consequences for miniaturized mechanical tests. *Materials Science and Engineering: A*, 459(1-2):262 – 272, 2007.

[130] M. G. Nicholas and C. F. Old. Liquid metal embrittlement. *Journal of Materials Science*, 14(1):1–18, 1979.

[131] W. Sigle, G. Richter, M. Ruhle, and S. Schmidt. Insight into the atomic-scale mechanism of liquid metal embrittlement. *Applied Physics Letters*, 89(12):121911, September 2006.

[132] Y. M. Wang, K. Wang, D. Pan, K. Lu, K. J. Hemker, and E. Ma. Microsample tensile testing of nanocrystalline copper. *Scripta Materialia*, 48(12):1581–1586, June 2003.

[133] R. O. Murty K. L. Horton J.A. Koch C. C. Youssef, K. M. Scattergood. Ultra-high strength and high ductility of bulk nanocrystalline copper. *Applied Physics Letters*, 87:1 –3, 2005.

[134] D. Gianola and C. Eberl. Micro- and nanoscale tensile testing of materials, March 2009.

[135] M. Jin, A.M. Minor, E.A. Stach, and J.W. Morris Jr. Direct observation of deformation-induced grain growth during the nanoindentation of ultrafine-grained Al at room temperature. *Acta Materialia*, 52(18):5381 – 5387, 2004.

[136] J.W. Morris, M. Jin, and A.M. Minor. In situ studies of the transmission of strain across grain boundaries. *Materials Science and Engineering: A*, 462(1-2):412 – 417, 2007. International Symposium on Physics of Materials, 2005.

[137] A.M. Hodge, Y.M. Wang, T.W. Barbee, and Jr. Large-scale production of nano-twinned, ultrafine-grained copper. *Materials Science and Engineering: A*, 429(1-2):272 – 276, 2006.

[138] Christoph Eberl. *digital image correlation.* http://www.mathworks.com/matlabcentral/fileexchange/12413, August 2010.

[139] P. A. Gruber, J. Böhm, F. Onuseit, A. Wanner, R. Spolenak, and E. Arzt. Size effects on yield strength and strain hardening for ultra-thin Cu films with and without passivation: A study by synchrotron and bulge test techniques. *Acta Materialia*, 56(10):2318–2335, June 2008.

[140] P.G. Sanders, J.A. Eastman, and J.R. Weertman. Elastic and tensile behavior of nanocrystalline copper and palladium. *Acta Materialia*, 45(10):4019–4025, October 1997.

[141] Y. Wang, M. Chen, F. Zhou, and E. Ma. High tensile ductility in a nanostructured metal. *Nature*, 419(6910):912–915, October 2002.

[142] D. Yu and F. Spaepen. The yield strength of thin copper films on kapton. *Journal of Applied Physics*, 95, 2004.

[143] O. Kraft, M. Hommel, and E. Arzt. X-ray diffraction as a tool to study the mechanical behaviour of thin films. *Materials Science and Engineering: A*, 288(2):209 – 216, September 2000.

[144] W. Nix. Mechanical properties of thin films. *Metallurgical and Materials Transactions A*, 20(11):1989, 1989-11-01.

[145] P. A. Gruber. Mechanical properties of ultra thin metallic films revealed by synchrotron techniques. Ph.D. thesis, Universität Stuttgart, 2007.

[146] H. Huang and F. Spapen. Tensile testing of free-standing Cu, Ag, and Al thin films and Ag/Cu multilayers. *Acta Materialia*, 48:3261–3269, 2000.

[147] D.T. Read. Tension-tension fatigue of copper thin films. *Int. J. Fatigue*, 20:203–209, 1998.

[148] R.R Keller, J.M Phelps, and D.T Read. Tensile and fracture behavior of free-standing copper films. *Materials Science and Engineering: A*, 214:42–52, August 1996.

[149] W. D. Nieman, J. Weertman, and R. W. Siegel. Mechanical behavior of nanocrystalline Cu and Pd. *J. Mater. Res.*, 6:1012 – 1027, 1991.

[150] D. Read. Young's modulus of thin films by speckle interferometry. *Meas. Sci. Technol.*, p, 1998.

[151] D. V. Bachurin and P. Gumbsch. Elastic and plastic anisotropy after straining of nanocrystalline palladium. *Phys. Rev. B*, 85(8):085407–, February 2012.

[152] J. Weissmüller, J. Markmann, M. Grewer, and R. Birringer. Kinematics of polycrystal deformation by grain boundary sliding. *Acta Materialia*, 59(11):4366–4377, June 2011.

[153] R.K. Nalla, J.H. Kinney, and R.O. Ritchie. Mechanistic fracture criteria for the failure of human cortical bone. *Nat Mater*, 2(3):164–168, March 2003.

[154] D.A.W. Kaute, H.R. Shercliff, and M.F. Ashby. Delamination, fibre bridging and toughness of ceramic matrix composites. *Acta Metallurgica et Materialia*, 41(7):1959–1970, July 1993.

[155] G. Bao and Z Zuo. Remarks on crack-bridging concepts. *Appl Mech Rev*, 45:355–366, 1992.

[156] Seong-Woong Kim, Xiaoyan Li, Huajian Gao, and Sharvan Kumar. In situ observations of crack arrest and bridging by nanoscale twins in copper thin films. *Acta Materialia*, (0):in press, 2012.

[157] ASTM 399-90. Standard test method for plane-strain fracture toughness of metallic materials. *ASTM International*, 1997.

[158] L. Lu, R. Schwaiger, Z.W. Shan, M. Dao, K. Lu, and S. Suresh. Nano-sized twins induce high rate sensitivity of flow stress in pure copper. *Acta Materialia*, 53(7):2169 – 2179, 2005.

[159] Y.F. Shen, L. Lu, Q.H. Lu, Z.H. Jin, and K. Lu. Tensile properties of copper with nano-scale twins. *Scripta Materialia*, 52(10):989–994, May 2005.

[160] X. Zhang, H. Wang, X. H. Chen, L. Lu, K. Lu, R. G. Hoagland, and A. Misra. High-strength sputter-deposited Cu foils with preferred orientation of nanoscale growth twins. *Applied Physics Letters*, 88(17):173116 1–3, April 2006.

[161] N. Li, J. Wang, J.Y. Huang, A. Misra, and X. Zhang. Influence of slip transmission on the migration of incoherent twin boundaries in epitaxial nanotwinned Cu. *Scripta Materialia*, 64(2):149–152, January 2011.

[162] A.M. Hodge, Y.M. Wang, and T.W. Barbee Jr. Mechanical deformation of high-purity sputter-deposited nano-twinned copper. *Scripta Materialia*, 59(2):163–166, July 2008.

[163] S. Xie T.W. Barbee Jr. A.M. Hodge C.J. Shute, B.D. Myers and J.R. Weertman. Microstructural stability during cyclic loading of multilayer copper/copper samples with nanoscale twinning. *Scripta Materialia*, 60:1073–1077, 2009.

[164] C.J. Shute, B.D. Myers, S. Xie, S.-Y. Li, T.W. Barbee Jr., A.M. Hodge, and J.R. Weertman. Detwinning, damage and crack initiation during cyclic loading of Cu samples containing aligned nanotwins. *Acta Materialia*, 59(11):4569–4577, June 2011.

[165] J. Wang, N. Li, O. Anderoglu, X. Zhang, A. Misra, J.Y. Huang, and J.P. Hirth. Detwinning mechanisms for growth twins in face-centered cubic metals. *Acta Materialia*, 58(6):2262–2270, April 2010.

[166] D. LaVan and W. Sharpe. Tensile testing of microsamples. *Experimental Mechanics*, 39(3):210–216, September 1999.

[167] C. Mattheck. *Thinking Tools After Nature*. 2011.

[168] Andreas Kienzler. Auswirkungen mechanischer Oberflächenbehandlungen auf die Randschichteigenschaften mikrostrukturierten Formeinsätze. PhD thesis, Karlsruhe Insitute of Technology, 2010

[169] J. J. Fundenberger, A. Morawiec, E. Bouzy, and J. S. Lecomte. Polycrystal orientation maps from TEM. *Ultramicroscopy*, 96(2):127–137, August 2003.

[170] H. H. Liu, S. Schmidt, H. F. Poulsen, A. Godfrey, Z. Q. Liu, J. A. Sharon, and X. Huang. Three-dimensional orientation mapping in the transmission electron microscope. *Science*, 332(6031):833–834, May 2011.

[171] J. Prokop, C. Eberl, M. Funk, P. Prüfe, J. Lorenz, M. Piotter, M. Welz, and Ritzhaupt-Kleissl. Mechanical testing of micro samples produced by a novel LIGA related process chain. *Microsyst Technol*, 17:281–288, 2011.

[172] L.-C. Lai, W.-A. Chiou, and J.C. Earthman. Influence of electrical discharged machining and surface defects on the fatigue strength of electrodeposited nanocrystalline Ni. *International Journal of Fatigue*, 32(3):584–591, March 2010.

[173] Edward C. Courtney. *Mechanical Behavior of Materials*. Waveland Press, Inc., 2000.

[174] H.W. Hoeppel, M. Brunnbauer, Mughrabi H., R. Z. Valiev, and A.P. Zhilyaev. Cyclic deformation behaviour of ultrafine grain size copper produced by equal channel angular extrusion. *Proc of Werkstoffwoche 2000*, 2001.

[175] A. Vinogradov and S. Hashimoto. Multiscale phenomena in fatigue of ultra-fine grain materials : an overview. *Mater Trans*, 42:74 – 83, 2001.

[176] F.-L. Liang and C. Laird. The effect of environment on the mechanism of fatigue crack initiation and propagation in polycrystalline copper. *Materials Science and Engineering: A*, 117(0):83–93, September 1989.

[177] L. Lu, M. Dao, T. Zhu, and J. Li. Size dependence of rate-controlling deformation mechanisms in nanotwinned copper. *Scripta Materialia*, 60(12):1062–1066, June 2009.

[178] Y.F. Shen, L. Lu, M. Dao, and S. Suresh. Strain rate sensitivity of Cu with nanoscale twins. *Scripta Materialia*, 55(4):319–322, August 2006.

[179] W.S. Zhao, N.R. Tao, J.Y. Guo, Q.H. Lu, and K. Lu. High density nano-scale twins in Cu induced by dynamic plastic deformation. *Scripta Materialia*, 53(6):745–749, September 2005.

[180] R. Z. Valiev. Structure and mechanical properties of ultrafine-grained metals. *Materials Science and Engineering: A*, 234-236(0):59–66, August 1997.

[181] J.A. Brown and N.M. Ghoniem. Structure and motion of junctions between coherent and incoherent twin boundaries in copper. *Acta Materialia*, 57(15):4454–4462, September 2009.

[182] Y. Wei. The kinetics and energetics of dislocation mediated de-twinning in nano-twinned face-centered cubic metals. *Materials Science and Engineering: A*, 528(3):1558–1566, January 2011.

[183] M. Chen, E. Ma, K.J. Hemker, H. Sheng, Y. Wang, and X. Cheng. Deformation twinning in nanocrystalline aluminum. *Science*, 300(5623):1275–1277, May 2003.

[184] Z.S. You, L. Lu, and K. Lu. Tensile behavior of columnar grained cu with preferentially oriented nanoscale twins. *Acta Materialia*, 59(18):6927–6937, October 2011.

[185] J. Weertman and j.R. Weertman. *Elementary Dislocationn Theory*. New York: Oxford University Press, 1992.

[186] Y. Xiang, X. Chen, and J. Vlassak. Plane-strain bulge test for thin films. *J. Mater. Res.*, 20:2360 – 2370, 2005.

[187] J. Vlassak and W. Nix. A new bulge test technique for the determination of young's modulus and poisson's ratio of thin films. *J. Mater. Res.*, 7:3242 – 3249, 1992.

[188] W. Oliver and G. Pharr. An improved technique for determining hardness and elastic modulus using load and displacement sensing indentation experiments. *J. Mater. Res.*, 7:1564–1583, 1992.

[189] T. Malis, S.C. Cheng, and R.F. Egerton. Eels log-ratio technique for specimen-thickness measurement in the TEM. *Journal of Electron Microscopy Technique*, 8:193–200, 1988.

[190] L. Reimer. Transmission Electron Microscopy. Spinger-Verlag, 1984.

[191] D.B. Williams and C.B. Carrter. *Transmission Electron Microscopy*. Plenum Press, New York, 1996.

[192] J. Wang, A. Misra, and J. P. Hirth. Shear response of $\Sigma 3$ {112} twin boundaries in face-centered-cubic metals. *Phys. Rev. B*, 83(6):064106–, February 2011.

[193] Guillaume Wiederhirn. The strength limits of ultra-thin copper films. PhD thesis, Max-Plank-Institut für Metallforschung, Stuttgart, 2007.

[194] J. Wang, O. Anderoglu, J. Hirth, A. Misra, and X. Zhang. Dislocation structures of sigma 3 {112} twin boundaries in face centered cubic metals. *Applied Physics Letters*, 95, 2009.

[195] L. Xu, D. Xu, K.N. Tu, Y. Cai, N. Wang, P. Dixit, J. Pang, and J. Miao. Structure and migration of (112) step on (111) twin boundaries in nanocrystalline copper. *Journal of Applied Physics*, 104, 2008.

[196] S. V. Divinski. Diffusion in nanostructured materials. *Defect and Diffusion Forum*, pages 623–632, 2009.

[197] T. J. Rupert, D. S. Gianola, Y. Gan, and K. J. Hemker. Experimental observations of stress-driven grain boundary migration. *Science*, 326(5960):1686–1690, December 2009.

[198] J. Schiotz. Strain-induced coarsening in nanocrystalline metals under cyclic deformation. *Materials Science and Engineering: A*, 375:975–979, July 2004.

[199] I.A. Ovid'ko, A.G. Sheinerman, and E.C. Aifantis. Stress-driven migration of grain boundaries and fracture processes in nanocrystalline ceramics and metals. *Acta Materialia*, 56(12):2718–2727, July 2008.

[200] J. W. Cahn and J. E. Taylor. A unified approach to motion of grain boundaries, relative tangential translation along grain boundaries, and grain rotation. *Acta Materialia*, 52(16):4887–4898, September 2004.

[201] M. Yu. Gutkin and Ovid'ko I.A. Grain boundary migration as rotational deformation mode in nanocrystalline materials. *Applied Physics Letters*, 87, 2005.

[202] AP. Sutton and RW. Ballufi. Interfaces in crystalline materials. Oxford: Oxford University Press, 1995.

[203] G. Richter, K. Hillerich, D. S. Gianola, R. Mönig, O. Kraft, and C. A. Volkert. Ultrahigh strength single crystalline nanowhiskers grown by physical vapor deposition. *Nano Lett.*, 9(8):3048–3052, July 2009.

Schriftenreihe
des Instituts für Angewandte Materialien

ISSN 2192-9963

Die Bände sind unter www.ksp.kit.edu als PDF frei verfügbar oder
als Druckausgabe bestellbar.

Band 1   Prachai Norajitra
           Divertor Development for a Future Fusion Power Plant. 2011
           ISBN 978-3-86644-738-7

Band 2   Jürgen Prokop
           **Entwicklung von Spritzgießsonderverfahren zur Herstellung von
           Mikrobauteilen durch galvanische Replikation.** 2011
           ISBN 978-3-86644-755-4

Band 3   Theo Fett
           **New contributions to R-curves and bridging stresses – Applications
           of weight functions.** 2012
           ISBN 978-3-86644-836-0

Band 4   Jérôme Acker
           **Einfluss des Alkali/Niob-Verhältnisses und der Kupferdotierung auf
           das Sinterverhalten, die Strukturbildung und die Mikrostruktur von
           bleifreier Piezokeramik $(K_{0,5}Na_{0,5})NbO_3$.** 2012
           ISBN 978-3-86644-867-4

Band 5   Holger Schwaab
           **Nichtlineare Modellierung von Ferroelektrika unter Berücksich-
           tigung der elektrischen Leitfähigkeit.** 2012
           ISBN 978-3-86644-869-8

Band 6   Christian Dethloff
           **Modeling of Helium Bubble Nucleation and Growth in Neutron
           Irradiated RAFM Steels.** 2012
           ISBN 978-3-86644-901-5

Band 7   Jens Reiser
           **Duktilisierung von Wolfram. Synthese, Analyse und Charakterisierung
           von Wolframlaminaten aus Wolframfolie.** 2012
           ISBN 978-3-86644-902-2

Band 8   Andreas Sedlmayr
           Experimental Investigations of Deformation Pathways in Nanowires.
           2012
           ISBN 978-3-86644-905-3

**Schriftenreihe
des Instituts für Angewandte Materialien**

ISSN 2192-9963

Band 9   Matthias Funk
          **Microstructural stability of nanostructured fcc metals during cyclic
          deformation and fatigue.** 2012
          ISBN 978-3-86644-918-3